**BEIJING ANCIENT ARCHITECTURE SERIES**

# TEMPLES AND MOSQUES

Compiled by Beijing Ancient Architecture Research Institute

Beijing Publishing Group
Beijing Arts and Photography Publishing House

图书在版编目（CIP）数据

寺观：英文 / 北京市古代建筑研究所编. — 北京：北京美术摄影出版社，2015.5
（北京古建文化丛书）
ISBN 978-7-80501-751-8

Ⅰ. ①寺… Ⅱ. ①北… Ⅲ. ①寺庙—古建筑—建筑艺术—北京市—英文 Ⅳ. ①TU-092.2

中国版本图书馆CIP数据核字（2015）第049490号

Planner: Li Qingxia

Executors: Dong Weidong, Qian Ying

Executive Editor: Huang Wenwen

Translators: Sun Libing, Zhou Chunhong

Production Supervisor: Peng Junfang

BEIJING ANCIENT ARCHITECTURE SERIES

## TEMPLES AND MOSQUES

SI GUAN

**Compiled by Beijing Ancient Architecture Research Institute**

Publisher: Beijing Publishing Group
     Beijing Arts and Photography Publishing House

Address: No. 6, Beisanhuan Zhonglu, Beijing

Zip Code: 100120

Website: www.bph.com.cn

Distributor: Bpg Art Media (Beijing) Co., Ltd

Dealer: Xinhua Bookstore

Printer: Beijing Chengye Hengxin Printing Co., Ltd

Edition: May, 2015, 1st edition

Format: 889 mm by 1194 mm 1/16

Printed Sheet: 19.5

Words: 81, 000

Book No.: ISBN 978-7-80501-751-8

Price: RMB 236.00 (yuan)

Tel: 010-58572393

# Editorial Board

**Chief Editor:**    Han Yang

**Associate Editor:**  Hou Zhaonian

**Editorial Board Members:**

Wang Xia, Li Weiwei, Shen Yuchen, Pang Yong, Hou Zhaonian,

Jiang Ling, Dong Liang, Han Yang

**Compilers:**    Li Weiwei, Shen Yuchen, Dong Liang, Jiang Ling, Wang Xia

Pang Yong

**Photographers:**  Wang Song, Wang Zhen, Wang Jing, Wang Jianhua, Wang Hongkun,

Wang Hongjie, Niu Dingyin, Deng Jingru, Dai Zhigang, Liu Fengzhen,

Qi Honghe, Qi Qingguo, Du Dianwen, Li Weiwei, Li Ziqiang,

Li Zhendong, Yang Jiyun, Wu Jianhua, He Bingfu, Zhang Chengzhi,

Zhang Aimin, Zhang Zhaoji, Pang Yong, Hu Dunzhi, Jiang Nan,

Jiang Jingyu, Xuan Lipin, Yao Tianxin, Qian Ying, Xu Zifeng,

Xu Danyu, Gao Mei, Liang Yugui, Liang Zhihui, Fu Gongyue,

Wei Gang, etc.

**Drawers:**    Ma Yuyang, Wang Wei, Liu Jia, He Zhimin, Zhang Jingyang,

Pang Yong, Zhao Jinjun, Han Yang, Li Dongqing

# PREFACE

More than 3,000 years of city history and over 800 years of capital history have left this land of Beijing with 3,500-odd cultural heritages. Among them are the Great Wall winding its way among the mountains, the ancient villages with smoke spiraling from kitchen chimneys, the palace where emperor decreed and lived, the garden where they enjoyed their leisure time, the mausoleums where they were buried after death, the imperial ancestral temples and sacrificial altars, the temples or Taoist temples striking the bell at dawn and playing the drum at dusk, the magnificent princely mansions, the tranquil civilian houses, the inscribed steles silent in brambles and the Buddhist pagodas towering into the clouds. They bestowed on Beijing profound culture and traditions.

In recent years, with the development of social civilization, the preservation of cultural heritages has been drawing more and more attentions from all walks of life. And this cause has been furthered unprecedentedly. Most of the historic buildings have got effective protection in Beijing, a major city of cultural heritage. To better display and publicize these valuable heritages characteristic of rich cultural connotations, the compilation of the *Beijing Ancient Architecture Series* was organized by the Beijing Ancient Architecture Research Institute.

These books would be a record of existing excellent historic architecture in Beijing by words, pictures and drawing paper. We hope in this way we could display the art treasure, which focusing the intelligence and wisdom of the ancients, has witnessed Beijing's rise and fall during a thousand years of development.

Under the consideration of architecture, this series of books are to explore historic buildings from the point of view of cultural heritage and art. They aim at igniting the passion in readers to love these historic heritages abounding in cultural traditions and boundless artistic charm.

These books total ten volumes. Each focuses on one type of the ten categories of architectural heritages showing different functions and styles. All of them include the most typical and distinctive examples. Thus they would show readers the features of each category and detailed traits of each building as well as the whole appearance of Beijing historic architecture.

In total, this series of books have over one million words, more than 4,000 photos and several hundreds of ink drawings. On one hand, they are continuation of predecessors' experiences. On the other hand, they are the summary of our work over the years.

Beijing Ancient Architecture Research Institute

# GUIDE TO THE USE OF THIS BOOK

1. This book is included as one of the volumes of the *Beijing Ancient Architecture Series*, it mainly introduces the temple constructions in Beijing area.

2. Dozens of representative and typical heritage conservation units were selected from about hundreds of temple constructions existent in Beijing and were recorded in this volume.

3. This book manages to show and interpret the characteristics of the architectural art and the cultural connotations roundly from the macro and micro perspectives by combining words with drawings and photos. It also veritably records the ancient temple constructions preserved and existent in Beijing.

4. The content of this book includes the outline of the ancient temples in Beijing and the examples of temple constructions existent in Beijing.

5. The section with examples of temple construction is categorized into Buddhist architecture (Han Buddhism and Tibetan Buddhism), Taoist architecture and Islamic architecture. The same category was also subdivided into several small categories and sorted out first according to the time sequence and then according to the rank from high to low.

6. In July, 2010, Beijing Municipal Government adjusted the administrative divisions of its core functional areas. Since these series of books had been finalized and ready for printing at the time, the names of the administrative regions of Beijing used in these books are still those before the adjustment.

# CONTENTS

| | |
|---|---|
| TEMPLES AND MOSQUES IN BEIJING | 001 |
| Buddhist Temples | 001 |
| **Han Buddhist Temples** | 004 |
| The Tanzhe Temple | 004 |
| The Yunju Temple | 016 |
| The Jietai Temple | 023 |
| The Shifang Pujue Temple | 031 |
| The Fayuan Temple | 038 |
| The Dajue Temple | 043 |
| The Badachu Park on the Western Hills | 049 |
| The Shangfang Mountain Temples and the Yunshui Cave | 064 |
| The Cishan Temple | 068 |
| The Biyun Temple | 073 |
| The Wanshou Temple | 083 |
| The Baoguo Temple | 091 |
| The Guangji Temple | 094 |
| The Bailin Temple | 098 |
| The Fahai Temple | 103 |
| The Zhihua Temple | 107 |
| The Hongluo Temple | 113 |
| The Cheng'en Temple | 116 |
| The Guanghua Temple | 121 |
| The Dahui Temple | 125 |

| | |
|---|---|
| The Mohe Nunnery | 128 |
| The Changchun Temple | 132 |
| The Juesheng Temple (The Big Bell Temple) | 134 |
| The Lingyue Temple | 139 |
| The Main Hall of the Lingyan Temple | 142 |
| The Relics of the Cross Temple | 144 |
| The Tiewa Temple | 146 |

## Tibetan Buddhist Temples 148

| | |
|---|---|
| The Miaoying Temple | 148 |
| The Yonghegong Lama Temple | 155 |
| The Songzhu Temple and the Zhizhu Temple | 172 |
| The Main Hall of the Pudu Temple | 178 |
| The Fuyou Temple | 180 |

## Taoist Temples 185

| | |
|---|---|
| The White Cloud Temple | 188 |
| The Dagaoxuan Temple | 202 |
| The Beijing Dongyue Temple | 210 |
| The Guangren Palace | 222 |
| The Beiding Goddess Temple | 225 |
| The Miaofeng Hill Goddess Temple and Lingguan Palace | 228 |
| The Relics of the Yaji Hill Bixia Yuanjun Temple | 231 |
| The Fire God Taoist Temple | 236 |

The Xuanren Temple 242

The Ninghe Temple 244

The Sanjiadian Dragon King Temple 246

## Islamic Mosques 251

The Niujie Mosque 254

The Tongzhou Mosque 262

The Dongsi Mosque 266

The Huashi Mosque 272

POSTSCRIPT 275

# TEMPLES AND MOSQUES IN BEIJING

Beijing was an ancient city in Shang and Zhou Dynasties, an important settlement in Yan mountain areas and served as capital for five dynasties. With the development and spread of taoism in China, the Taoist temples, as a medium and a site to spread Taoist culture, have experienced a process from being small to being large, from being prosperous to being declining.

## 1 Introduction of the Development of the Temples and Mosques in Beijing

### 1.1 Introduction of the Temple and Mosque Architecture from the Three Kingdoms Period, Jin and Northern Dynasties to the Tang and Five Dynasties

During the Northern Wei Dynasty, Youzhou was one of the areas where Buddhism was prosperous. Men of letters and with literary reputation often made friends with the monks. Lu Jingyu from Fanyang area was a typical example. Being a Confucius scholar, he knew a lot about Buddhism. "Whenever the Indian Samana speaks of the scripture, they usually ask Jingyu to write a preface to it." Most of scholars would also like to talk about Buddhist theories, which made the culture of Youzhou area have a Buddhist feature. At that time, the Buddhist temples sprang up in Youzhou and delicate statues were set up there too. *Shuijing Zhu—Lei River* records: "Lei River (today's Yongding River) passes by Lingnan of Yan mountain, where there is an underground road whose northwestern end leads to Ji County. Emperor Xuanwu of Northern Wei Dynasty built stupas and temples and dug a tunnel under the ground. This road is exclusive only to the royal family and nobody can find it." The oldest temple in Beijing, according to historical

▲ Main Gate of the Tanzhe Temple

literature, is Tanzhe Temple in Beijing which was founded in the 1st year under the reign of Emperor Yongjia in Western Jin Dynasty. Temples at that time of a relatively large scale also included the Guanglin Temple in Northern Wei Dynasty and the Zhiquan Temple in Eastern Wei Dynasty.

With the spreading of Buddhism in Youzhou, Beijing had become an important area where Buddhism flourished until the period of Sui and Tang Dynasty. The eminent monks and Buddhist temples came forth in large numbers. A good many famous temples were built such

▲ Wangong Tower in the Yunju Temple

as the Yunju Temple, the Minzhong Temple (today's Fayuan Temple), the Huiju Temple (today's Jietai Temple) and the Yuni Temple (later called the Jiufeng Temple), among which, the Yunju Temple is worth mentioning for the scriptures inscribed on stones. Learning a lesson from the event of "Two Emperors surnamed Wu prohibiting Buddhism" (the two emperors were Emperor Taiwu in Northern Wei Dynasty and Emperor Zhouwu in Northern Zhou Dynasty), Master Jingwan in Sui and Tang Dynasties engraved the scriptures on the stone boards so that they could be preserved forever. Though there were not any stone scriptures at that time, the high quality stone scriptures needed technological and economic support, which showed that Buddhism had become more and more popular in Beijing at that time. The engraved scriptures have been preserved as long as a thousand years since then. In the middle of Tang Dynasty, due to the popularity of Buddhism in Beijing area, a large number of well-known temples were built, such as the Heavenly Kings' Temple (today's Tianning Temple), the Guiyi Temple, the Youtang Temple, the Zhenying Temple, the Chongxiao Temple, the Jingye Temple, etc. Buddhism had affected central governments' financial income and social safety. For instance, monks who possessed a large amount of real estate did not pay tax; there were too many monks and not all of them behaved well. As a result, Buddhism experienced a disaster in the middle of Tang Dynasty when emperor Wuzong commanded to destroy Buddhism. "Only in the areas of Zhen, You, Wei and Lu on the north of Yellow River were Buddhism always respected—no temples were demolished and no monks were dismissed to secular life." Beijing was in Youzhou area, so

▲ Hall of the Heavenly Kings in the Fayuan Temple

▲ Jietan Hall in the Jietai Temple

Buddhism was not greatly affected in this disaster. In late Tang Dynasty, the trend of building Buddhism temples were more and more evident. Temples such as the Shengguo Temple, the Baoji Temple, the Yousheng Temple, the Xingshan Temple, the Shaosheng Temple and the Xiangshui Temple, etc were built. The wide spreading of Buddhism in Tang Dynasty laid a foundation for its further popularity in Beijing in Liao and Jin Dynasties.

In that period, the indigenous Taoism developed well along with Buddhism in Beijing. The two religions blended with each other and the famous Taoism temple— the White Cloud Temple known as the first temple in the north was built at that time. At the end of Tang Dynasty and the beginning of Five Dynasties, Taoism began to develop. The local separatist regime leader named Liu Rengong in the late Tang Dynasty was fond of alchemy. He not only was apprenticed to the Taoist Wang Ruona and learned alchemy and eclosion, but also built the Yuhe County in the western suburbs of Youzhou area for him. Taoism became more and more popular owing to the authorities' favors.

▲ Main Gate of the White Cloud Temple

## 1.2 Introduction of the Temple and Mosque Architecture in Liao and Jin Dynasties

The rulers in Liao Dynasty believed in Buddhism, and they strongly supported the development of Buddhism so that Buddhism in Liao was as popular as it was in Tang and much more popular than in Song Dynasty. In Liao Dynasty, Beijing was the auxiliary capital called Nanjing where Buddhist ceremonies prevailed. It took the first place in the prosperity of Buddhism among the five capitals. As *Khitan National History* records, "The number of monks in Buddhist temples of Yanjing is the largest in the north." According to the Journey Records in Songmo written by Hong Hao , there were 36 large temples and numerous small temples in Yanjing. During about a hundred years under the reign of Emperor Shengzong, Emperor Xingzong and Emperor Daozong of Liao Dynasties, the country came to its great prosperity, and Buddhism in Beijing also became more and more flourishing. There were a large number of Buddhist temples inside and outside Nanjing. The Minzhong Temple built in Tang Dynasty became the venue for religious, political and diplomatic activities. According to the *Dominant Yuan History*, the 5th year under the reign of Emperor Qingning (1059), Princess Qinyue in Liao made the Tang Yinfang Mansion as a temple, which was the Haotian Temple of Liao Dynasty and occupied almost a hundred hectares. Princess Qinyue had a distinguished identity: she was the daughter of Emperor Shengzong in Liao, the elder sister of Xingzong and the mother of Queen Yide. Master Miaoxing, who was responsible for the construction of the temple, was a Khitan and belonged to the clans of minister Chu Guowang who was the maternal uncle of the prince. Consequently, the Haotian Temple in Nanjing of Liao held a superior position than other temples. The Haotian Temple surpassed the Minzhong temple in splendor not in long history. Although the relics of the Haotian Temple can not be found now, it  was indeed a well-known large temple at that time. The temples recorded in literature of Nanjing in Liao also include the Guiyi Temple, the Xianlu Temple, the Tianwang Temple (today's Tianning Temple), the Yanshou Temple, the Xingguo Temple, the Miaoying Temple (today's Baita Temple), the Huiju Temple (today's Jietai Temple), the Qingshui Temple (today's Dajue Temple), etc. As one of the important policies of supporting

Buddhism, in the 7th year under the reign of Emperor Shengzong in Liao (1027), Emperor Shengzong resumed the scripture engraving activities in Yunju Temple of being which was suspended in late Tang Dynasty. During the years under the reign of Emperor Daozong, four Buddhist scriptures including *Nirvana*, *Kegon*, *Prajñā* and *Baoji* were finished and 47 slip-cases were engraved; during the 9th to 10th years under the reign of Emperor Da'an (1093-1094), 44 more slip-cases were engraved, containing over 6,000 engraved stones which become one of the most valuable treasures of the Yunju Temple and are preserved until now. These engraved scriptures make great contributions to the spreading and revisions of Buddhist scriptures.

There were many ethnic groups in Liao Dynasty, so there were also many kinds of religious beliefs. The rulers in Liao Dynasty believed in Buddhism, but they did not reject other religions and took the all-encompassing religious policy. As a result, the development and expansion of other religions were not at all blocked. In early Liao, Taoism and Buddhism were introduced simultaneously into Khitan. According to literature, among the various districts in Shangjing, the Youzhou and Jizhou districts had the largest number of Buddhist monks, nuns and Taoist priests, which showed that Taoism in Liao was introduced from Beijing area into other places. After Liao occupied Youzhou, the rulers did not reject Taoism. In Liao Dynasty, many aristocrats believed in both Buddhism and Taoism. Taking Emperor Shengzong and his brothers for example: Shengzong believed in Buddhism; one of his younger brothers Longyu believed in Taoism; the other younger brother was fond of martial arts. Yelv Longyu "has adored Taoism since childhood; whenever he met the Taoist priests, he was much pleased." According to the *History and Geography of Liao Dynasty,* "there were numerous markets, government offices and temples in Nanjing." The temples include the Tianchang Temple (today's White Cloud Temple), the Longxing Temple and the Yuan Temple, etc.

At the same time, Islam began to develop in Beijing. Many famous mosques such as the Niujie mosque were built.

After Zhongdu became the capital of Jin Dynasty, many emperors realized the negative influence of Buddhism on emperors in Liao who were obsessed with it and took a quite negative attitude to religion. In the 8th year under the reign of Emperor Shizong in Jin (1168), Shizong said to his courtiers: "I do not believe in Buddhism. Emperor Wu in Liang was enslaved by the Tongtai Temple, while Emperor Daozong in Liao presented household residence to the monks and appointed them government officials. Both of them were deeply misled." In the 19th year under his reign (1179), he said: "Most people believe in Buddhism and want to obtain happiness by chance. I was also misled in my early years, but finally found out that it could not be realized." Under the influence of such kinds of policies, most of the Buddhist temples in Jin Dynasty were built with the government supports or upon the imperial commands. The temples at that time include the Dasheng'an Temple, the Dajue Temple, the Hongfa Temple, the Dayong'an Temple, the Qingshou Temple (the Shuangta Temple in Yuan Dynasty), the Yangshan Xiyin Temple, the Bao'en Temple, the Fengfu Temple and the Fusheng Temple, etc.

However, Taoism which advocated temperance, physical and psychological edifying was attached great importance because its religious doctrines were in accordance with the rulers' requirements. "Jin adored Taoism. Just like Buddhism, ever since Taoism was introduced into Zhongzhou, it was soon introduced into the northern and southern parts of Yanjing. " At that time, a number of Taoist temples were built and rebuilt. In the early years under the reign of Emperor Shizong, the Tianchang Temple, which was ignored and abandoned in Liao Dynasty because of the popularity of Buddhism, was reconstructed and expanded in Jin Dynasty. The project lasted eight years and cost over 300 thousand strings of cash. The Yuxu Temple which was an old temple outside the city of Zhongdu was also reconstructed. After Yanjing was decided on as the capital of Jin Dynasty, the Yuxu Temple was rebuilt into the memorial temple for Wanyan Zongbi who was one of the founding fathers of the country. During the Taihe years under the reign of Emperor Zhangzong, it was resumed as the Taoism temple and the architecture inside the temple such as the Sanqing Hall was also rebuilt. In addition, many large temples such as the Xuanzhen Temple and the Wuhua Temple were also built by the government. Therefore, Taoism was advocated in Jin Dynasty. The scale of advocation was under certain control.

Owing to the support of the rulers, the religious temples and mosques attracted various prayers and developed gradually. Many well-known pagodas and stone pillars have been preserved until now, such as the Pagoda of Heavenly Kings Temple (today's Tianning Temple Pagoda) and the Yinshan Monk Tombs.

## 1.3 Introduction of the Temple and Mosque Architecture in Yuan and Ming Dynasties

During the Liao and Jin Dynasties, the Buddhist temples had considerable scale in Yanjing area. However, during the chaos caused by war in the late Jin, the temples inside Yanjing were almost demolished, and most of the monks moved to the south with Emperor Xuanzong. After the Mongolian army occupied Zhongdu, they were busy with the war and did not pay much attention to Buddhism so that the Buddhist buildings were completely destroyed in the war. "In the early years under the reign of Emperor Zhenyou, troops from heaven (Mongolian army) started a southern expedition and Yanjing soon surrendered. After the war, monks could be seen nowhere and were not sympathized by the government officials. All the temples were occupied by common people." Ruled by ethnical minorities, Yuan Dynasty respected Tibetan Buddhism greatly. In Beijing (which was called Dadu in Yuan Dynasty) which served as the capital of Yuan Dynasty, the Tibetan Buddhism developed rapidly. As early as Kublai was nominated to take charge of the affairs of Central Plains, he invited the leader of the Sariska School named Phagpa to come to Yanjing and put him into an important position. In the 1st year under the reign of Emperor Jingding (1260) when Kublai was enthroned, Kublai appointed him "Grand Precepto" (later called Imperial Precepto), "conferred him jade seal and made him in charge of Buddhism." Phagpa was promoted as the Karmapa after he created Mongolian new characters and became the highest leader of Buddhism in Yuan Dynasty. Until the end of Yuan, the highest leaders of Buddhism had always been the top leaders in Sariska School. Due to the support of the imperial court, the Tibetan Buddhism developed rapidly in Dadu and a group of high-level temples were built. The renowned temples inside and outside Dadu and in the suburban areas were: the Daxuan Wenhong Temple (inside today's Fragrant Hills Park), the Da Huguo Renwang Temple, the Da Shengshou Wan'an Temple (today's Miaoying Temple), the Da Tianshou Wanning Temple (in the east of today's Drum-tower), the Da chong'en Fuyuan Temple, the Da Chenghua Puqing Temple, the Dazhaoxiao Temple (today's Shifang Pujue Temple), the Dayongfu Temple, the Da Tianyuan Yansheng Temple (today's Zhengguo Temple in Badachu) and the Da Chengtian Husheng Temple. These ten temples has had a reputation for a long time and surpassed the Yanjing temples in Liao and Jin Dynasties in scale and the amount of properties. Besides, the White Pagoda which was the model building of Da Shengshou Wan'an Temple inside Dadu has been preserved until now.

As for Taoism, Genghis Khan summoned the head of Quanzhen School, Qiu Chuji Taoist priest in early Mongolian period, conferred him the title of Shenxian, Lord Grandmaster, and asked him in charge of Taoism in the country. After Qiu Chuji returned Yanjing, he first lived in the Tianchang Temple (today's White Cloud Temple) and later in the Yuxu Temple. He allowed Quanzhen School to build temples with liberty, so that "thousands of years has never seen the fast construction of Taoist temples." After Qiu Chuji passed away, his successors like Yin Zhiping continued to develop the Quanzhen School. During the 20 years from Ogodei to Mangu, Quanzhen Taoism continued to dominate other religions in Central Plains. However, because of its overlarge power, the fake Taoist priests were mixed with the genuine priests; because of mishandled relationship between Buddhism and Taoism and Confucianism, great contradictions were caused. When Kublai took charge of the political affairs in Central Plains, he took some measures to solve the contradictions between Taoism, Confucianism and Buddhism. The measures included supporting Confucianism and Buddhism and attacking Taoism. After he was enthroned, he took strict policies to repress various kinds of Taoism. As a result, Quanzhen School began to decline. But Zhengyi School which was another school of Taoism developed well in Dadu because one of his successors Zhang Liusun made good performance there so that the influence of Zhengyi School has reached as far as Dadu in the north. Zhang Liusun was entitled respectfully Zhang Shangqing by Emperor Kublai. Afterwards, Zhang Liusun's position became higher and higher. He and his successor Wu Quanjie built the Dongyue Rensheng Temple (today's Dongyue Temple) and reconstructed the Taiyi Yanfu Temple.

People in the Yuan Dynasty were divided into four classes. The top class were Mongolian people, the second class Semu people, the third class people of Han nationality, the four class Southern people of Han nationality. The Hui minority believing in Islam belongs to the second-level Semu people according to the ethnic policy in Yuan Dynasty. Many ministers served Kublai after he was enthroned were Semu people believing in Islam. In addition, a large number of Semu army, craftsmen, merchants, etc recruited from the Western Regions came to settle in Dadu, which promoted the development of Islam in Dadu. Islam, therefore became an important religion in Beijing. A large number of mosques were also built up. "There were over ten thousand mosques all over the country, as near as Beijing and as far as other cities. People went to the mosques to perform prayers, which resulted in other temples' emptiness."

In early Ming Dynasty, because Zhu Di thought that he was protected by Emperor Xuanwu in the fight for the throne, he started the trend of believing in Taoism in Ming Dynasty and most of the royal members in Ming Dynasty believed in Taoism. Emperor Jiajing also believed in and gave flattery to Taoism. He was obsessed by alchemy, so the Taoist priests around the country came in groups to try their luck. The Taoist architecture developed well in that period. The traditional Taoist temples such as the White Cloud Temple and Dongyue Temple attracted numerous believers. A number of famous Taoism temples were built in Beijing such as the Wuding Temple (Five Taishan Temples). What's more, in the 21st year under the reign of Emperor Jiajing (1542), the Dagaoxuan Hall which was used exclusively by royal family to worship Taoism was built and became the Taoist temple with the highest level now extant. Besides, because some of the Taoist gods were overlapped with those of the state rituals, some Taoist temples have become state ritual architectures since Ming Dynasty. What was worth mentioning was that the singers and dancers in the Tiantan Shenle Temple (Shenle Hall in Qing Dynasty) were all Taoist priests from Zhengyi School, which showed that Taoism played an important role in state rituals and proved that Taoism was at its height in Ming Dynasty.

▲ The Dagaoxuan Hall

### 1.4 Introduction of the Temple and Mosque Architecture in Qing Dynasty and the Period of the Republic of China

The rulers in Qing Dynasty advocated various kinds of religions, so all kinds of religions developed very well in that period. A large number of religious architecture appeared in Beijing. Among a few hutongs inside the city, there was a public religious architecture; in some districts, there were three to four religious architectures in just one hutong. According to the *Panoramic View of Beijing under Qianlong's Reign*, there were over 1,000 temples inside the city, small or large. In order to stabilize the borders and the whole country, the Qing regime used religion to govern the Mongolian and Tibetan regions. Therefore, a group of Tibetan Buddhist temples were built under imperial commands. These temples included the Yonghegong lama Temple, the Pudu Temple, the Huang Temple (the Donghuang Temple and the Xihuang Temple), etc. They were not only of large scale and high level, but also of delicate building techniques.

From late Qing Dynasty to the period of the Republic of China, with the

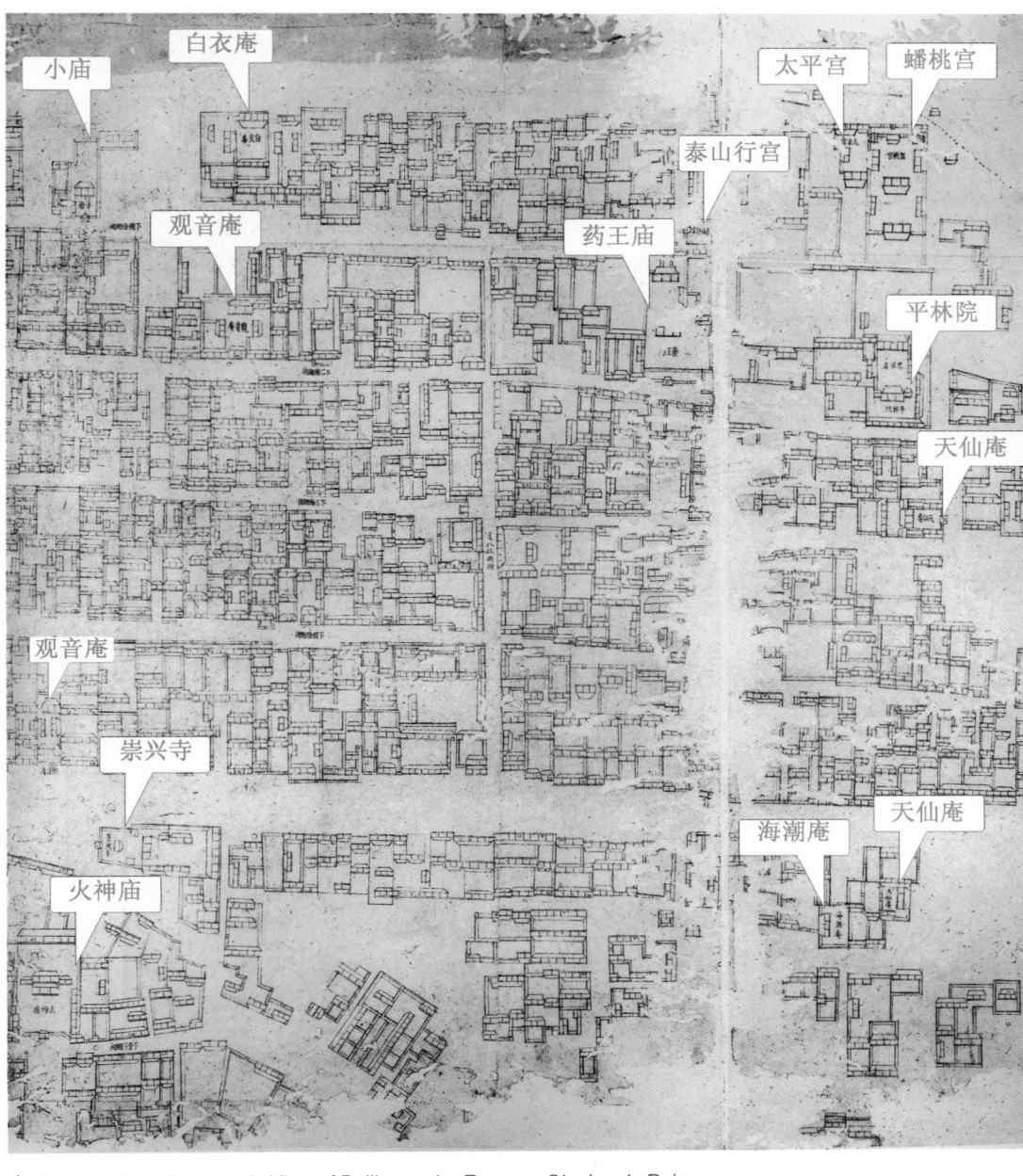

▲ Section of the *Panoramic View of Beijing under Emperor Qianlong's Reign*

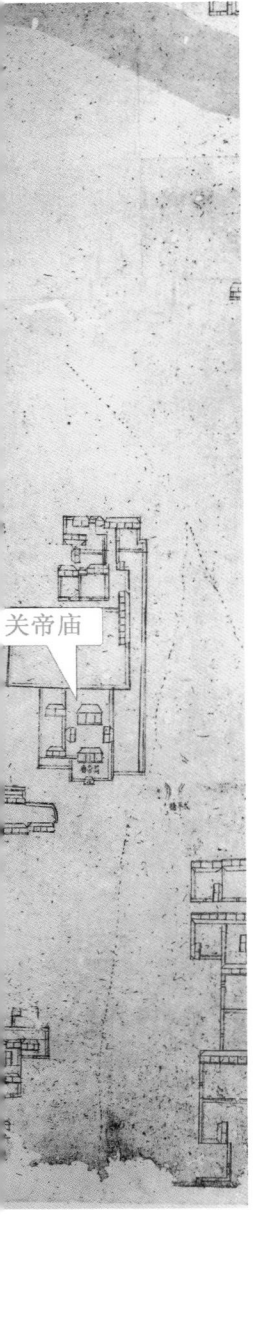

intrusion of western powers, the national power declined day by day. Except for a few western churches and nongovernmental temples, most temples began to decline for lack of renovation.

## 2  Introduction of the Temple and Mosque Architecture Currently Extant in Beijing

The temples currently extant in Beijing include the Buddhist temples, the Taoist temples and the Islamic mosques. These temples can be classified into large scaled and small scaled temples. The architectural layout of the large temples in Beijing usually consists of multi-roads with one axis in the middle (called central road). The main hall is usually situated in the central road, and the secondary halls and yards are mainly located symmetrically on the side roads. The buildings on each road consist of the main hall and the secondary halls which are laid out symmetrically and constitute the multi-row yard. The temples which are supported by government and royal families usually have large scale. They, especially those built upon imperial command are often of high level. The important halls in the temple usually have gable-and-hip roofs and hip roofs. The tiles used in the architecture are mostly large sized glazed tiles. Bracket sets, governmental tangent

▲ Mahavira Hall of the Lingyue Temple

circle color patterns and dragon seal color patterns are used to decorated the temples. The small temple is mostly nongovernmentally funded and the architecture level is relatively low. What's more, the glazed tiles are seldom used; the single building architecture usually has flush gable roofs and overhanging gable roofs. The temples can also be divided into temples on mountains and temples on plains. The single building architecture is mainly the traditional wooden framed hall. The temples were mainly built in Ming and Qing Dynasties bearing Ming and Qing architectural styles. The earliest temples preserved in Beijing are the hall of the Lingyue Temple and the Lingyan Temple of Yuan Dynasty architectural style.

## 3  Type and Feature of the Temple and Mosque Architecture Currently Extant in Beijing

These temples are a little different in terms of overall layout, single building architectural style and interior decoration because of their different religious doctrines, purpose for use, building time and sites.

### 3.1  Buddhist Temple Architecture

Chinese Buddhism has three schools: Han Buddhism (Mahayana Buddhism), Tibetan Buddhism and Theravada Buddhism (Sthavira). In Beijing, there are mainly Han Buddhism and Tibetan Buddhism, so the Buddhist temple architecture is mainly Han Buddhist architecture and Tibetan Buddhist architecture. The architecture includes temples, pagodas and stone statues. Temples are the focus of the present article.

#### 3.1.1  Han Buddhist Temples

There are over a hundred Han Buddhist temples extant in Beijing. The arrangement of hall follows the principle of "Buddha first, then Dhamma and Monk", meaning the Buddha hall being arranged in the middle, and the Dhamma and Monk halls on its sides. Among the fours building the Buddha hall is the most splendid. The Meditation hall and the Dhamma hall comes next. Then comes the monks' dormitories. In the forefront of the axis is the main gate (sometimes Pailou is built) which is also called Tri-gate and served as the entrance of the temple, often taking the form of the three-bay beamless hall. On both sides of the main gate are a bell tower and a drum tower (in the south- facing temples, the bell tower is in the east and the drum tower in the west, in accordance with the rule of the morning bell in the east and the evening drum in the west). In the face of the main gate is the Hall of the Heavenly Kings. Three halls are situated in a row. Behind the Hall of the Heavenly Kings is the second-row yard. Situated in the middle of the axis is the main hall, also named the Mahavira Hall. It is the gradest structure in the temple in terms of class and size. In the one-row yard or two-row yard behind the main hall is situated the secondary hall. In the rear yard is a two-story Pavilion for Storing Buddhist Scriptures (Sutra Pavilion) or Vairochana Pavilion. On both sides of axial yards are the meditation hall, the abstinence hall, the monk dormitories. The reason why the temples in Beijing usually have such layout is that they inherit the Tang and Song Dynasty's layout of Buddha halls in the front and Sutra pavilion in the back.

Although the temples in Beijing have the above mentioned similarities, they bear individual characteristics due to the different building time and sites.

Firstly, a long avenue ranging from several meters to several kilometers was built in front of most of the mountain temples in early times. The avenue not only makes the temple seem deeper, adding a leading way to the temple, bringing tranquility to the temple, but also forms a sharp contrast with the Buddhist halls and pavilions, highlighting the magnificence of the temple. Secondly, surrounded with the green pines and verdant cypresses before entering the temple, people would feel purified. For example, the leading avenue of the Tanzhe Temple is as long as several kilometers. The pagoda surrounded with trees (the Tanzhe Pagoda) makes the visitors feel calm and solemn. Early temples in Beijing such as the Jietai Temple, the Shifang Pujue Temple (the Wofo Temple), the Dajue Temple, etc have avenues too. The layout is similar to the layout of the mountain temples such as the Lingyin Temple, the Tiantong Temple, the Guoqing Temple, etc described in the "Five Mountains and Ten Temples" in Southern Song Dynasty. There are sayings such as "the Tiantong Temple with twenty li (one li = 0.5 kilometer) pine trees", "Guoqing Temple with ten li pine trees" and "Lingyin Temple with nine li pine path." Thirdly, some temples show

▲ Avenue in the Tanzhe Temple

the ethnic features. For example, temples such as the Dajue Temple and the Jietai Temple are all facing the east because these temples were built in Liao Dynasty and the east-facing layout shows Khitan clan's custom of facing the sun. Fourthly, one of the layout styles of the temples built on plains and in a city is the central position of the pagoda in the temples, such as the Tianning Temple. Lastly, there are other architectural layout features of the early temples. In Beijing, the veranda-yard layout around the main hall of the Shifang Pujue Temple that has been preserved ever since Sui and Tang Dynasties is quiet unique.

Although the temples in Ming and Qing Dynasties bear great similarities to temples in the previous dynasties, they have their own characteristics. Firstly, the royal temples, the temporary imperial abode are combined with the gardens. For example, the temporary imperial abodes in the Wanshou Temple, the Biyun Temple and the Tanzhe Temple. Secondly, they show the precious artistic features of the royal temples in Ming and Qing Dynasties. Typical examples are the fresco of Ming Dynasty in the Fahai Temple and the Cheng'en Temple, the color sculpture of Ming Dynasty in the Dahui Temple, the Yongle Bell of Ming Dynasty in the Big Bell Temple, "Beijing Music" and the Ten Thousand Buddha's statue in the Zhihua Temple and the dragon-type Tripitaka in the Bailin Temple. Besides, a large number of temples built in Ming and Qing Dynasties inherit and spread Buddhist culture with their different characteristics.

### 3.1.2 Tibetan Buddhist Temples

The architecture of Tibetan Buddhist temples is different from that of Han Buddhism temples due to their different religious doctrines and ethnic features. The Tibetan Buddhist temples appeared around Yuan Dynasty. An example is the famous Miaoying Temple in Beijing.

There are three types of Tibetan Buddhist temples. One is the traditional governmental wooden framework structure. This type of temple is featured by axisymmetry, courtyard-type layout and traditional governmental single-building hall with the decoration style, the Buddha enshrined and the functions of some hall reflecting Tibetan Buddhism. The famous Yonghegong Lama Temple in Beijing is an example. Another is the castle like structure of bricks and stones. This kind of temple is represented by the group of castle-like Tibetan temples in the rear hill of the Summer Palace and the Zhao Temple of the Fragrant Hills. The third is the combination of Han Buddhist temple hall with Tibetan pagoda. Generally, the front part of this kind of temple is traditional governmental wooden structure and the rear

▲ Adornments on the Roof of the Falun Hall in the Yonghegong Lama Temple

▲ Yizhu Xinjing Hall and White Pagoda in the Miaoying Temple

part is a tall white pagoda or a throne pagoda. The white pagoda and the diamond throne pagoda in Tibetan Buddhist temples have the same function as the stupa in Han Buddhist temple for the monks to chant scriptures around the pagoda or stupa. Temples of this kind in Beijing are the white pagoda in the Miaoying Temple, the Yong'an Temple in the Beihai park and the Xihuang Temple, etc. But what should be demonstrated is that not all the temples where white pagodas and throne pagodas can be seen are Tibetan Buddhist temples, because they have become an architectural style that have been used by Han Buddhist temples and Taoist temples.

Besides, the nature of the Tibetan Buddhist temples and Han Buddhist temples in Beijing don't stay the same. In some dynasties, the Tibetan Buddhist temple was transformed into the Han Buddhist temples, and vise versa.

## 3.2 The Taoist Temples and Other Temples' Architecture

There are two major Taoist schools in Beijing, Quanzhen School and Zhengyi School, whose architectural layout and the single building structure are basically the same with Han Buddhist temples. The only difference between Taoist and Buddhist architecture is that Taoist one follows the rule of the drum tower in the east and the bell tower in the west, while Buddhist one follows that of bell tower in the east and the drum tower in the west in accordance with the saying of the morning bell and the evening drum.

More than ten large Taoist temples are extant in Beijing. They consist of a temple gate, a main hall, a rear hall and some rear buildings, wing halls, the abbot residence and the abstinence halls on the two sides. The White Cloud Temple of Quanzhen School known as the First Temple in the north and Dongyue Temple of Zhengyi School known as the First Temple in the north now extant take such kind of layout. Besides, the well-known Wuding Temple, the Bixia Yuanjun Temple in the Miaofeng Hill and the Bixia Yuanjun Temple in the Yaji Hill are all very famous Taoist temples. Along with the Dongyue Temple, they constitute Dongyue gods' system of Beijing. The small Taoist temple is also of one-row or two-row yards. The number of small ones is too many to count. Because the Taoist gods' system is very complex and some of the Taoist gods are natural gods, some of the Taoist temples have the double identity of both Taoist temples and state ritual architecture. They are the sites for Taoist activities in ordinary days, and on the sacrifice days, they are the sites for the officials in charge of rituals sent by the state to offer sacrifices. According to *Qing Dynasty Record*, "When it was the time to offer sacrifices to the god of Beiji Yousheng Zhenjun each year, an official was sent to offer the sacrifices on Wanshou Festival. The god of Beiji Yousheng Zhenjun made his presence in the east of Rizhong Fang Bridge outside the Di'anmen…. When it was the time to offer sacrifices to the god of fire, the god of fire made his presence in Huoshen Temple in

▲ The Fire God Taoist Temple

the west of Rizhong Fang Bridge outside the Di'anmen.... When it was the time to offer sacrifices to the god of Dongyue each year, an official was sent to offer the sacrifices in the Dongyue Temple outside Chaoyangmen on Wanshou Festival." The overlapping situation is well shown by some Taoist temples where the gods and deified men are also enshrined. Those temples include the Zhenwu temple, the Fire God temple, the Dragon King temple and the Goddess Temple, etc . They are scattered across the cities and the villages.

What is worth mentioning is the mixture of Buddhism with Taoism. A number of Buddhist temples and Taoist temples on one hill constitute a large architectural complex and develop together forming the architectural treasure-house of both Buddhism and Taoism. The Shangfang Hill Temple group and the Miaofeng Hill architectural group are the two typical examples.

### 3.3 The Islamic Mosque Architecture

Islam was introduced to China in Tang Dynasty. The early Islamic mosques were completely of Arabian style. Later on, some mosques were of the traditional Chinese wooden framework structure, but they still showed the Islamic style in architectural system, overall layout and interior decorations.

Beijing is inhabited by various minority groups, and the Hui minority is one of them. Over 80 mosques are now extant in Beijing. They are

▲ Interior View of the Kiln Hall in the Niujie Mosque

▲ Brick Sculpture of Arabic Scripts of the Women Prayer Hall in the Niujie Mosque

▲ Interior View of the Prayer Hall in the Niujie Mosque

both similar with and different from the temples in terms of architectural layout and the single building hall. Firstly, on the layout, the mosque is still axisymmetrical and of courtyard-type, but they are facing the west where Mekka, the holy land of Islam is. It is different from the temple. Secondly, on the architectural elements, the Watching Moon Tower, the minaret and the bathroom are exclusive to the mosques. Thirdly, in terms of single-building structure and interior decorations, the architectural style is basically court-type, but the prayer hall which is the main hall is usually of the multi-floor type so that more rooms can be obtained. This is to meet the requirements of accommodating more people to perform the prayer. The main hall of the temples is usually of the one-floor style. Finally, on the decoration style, the mosque is decorated with the patterns of its own ethnic features.

Li Weiwei

There are many Han Buddhist temples. They can be classified into six categories based on their characteristics. The first category includes six early historic Buddhist temples represented by the Tanzhe Temple. The second category centers on three temple complexes that combine Buddhism, Taoism and folk beliefs and are featured by the Badachu Park on the Western Hills. The third category concentrates on the temples that combine temple constructions and the imperial temporary dwelling places, which are typified by the Biyun Temple and the Wanshou Temple. The Forth category introduces 12 temple constructions and artistic treasuries represented by the Fahai Temple. Represented by the Lingyue Temple and the Main Hall of the Lingyan Temple, the fifth category presents the temples that preserve the early constructions. Represented by the Cross Temple and the Iron Tile Temple, the sixth category includes the temples that mirror the long history of Nestorianism spread in Beijing and the architectural features of the temples in suburban Beijing.

There are also many Tibetan Buddhist temples in Beijing. This chapter introduces some representative Tibetan Buddhist temples in Beijing by selecting the most ancient Miaoying Temple, the Yonghegong Lama Temple that is used by the Qing government to administer affairs related to Tibetan Buddhism and the Pudu Temple that is renowned for its Manchu architectural features, etc.

## Buddhist Temples

## Han Buddhist Temples

## The Tanzhe Temple

Originally named the Xiuyun Temple, the Tanzhe Temple is located at the foot of the Tanzhe Mountain in the southeast of Mentougou District, Beijing. It is the most ancient temple dating back in history and the area the temple covers ranks the largest among all the temples in Beijing. It was listed by the State Council as a major historical and cultural site under state protection in 2001.

Constructed in the 1st year (307) during the reign of Emperor Yongjia in the Western Jin Dynasty, the Tanzhe Temple was originally named Jiafu Temple or Temple of Auspicious Fortune and has a history of over 1700 years. It was renamed the Dragon Spring Temple in the Tang Dynasty and was expanded and renovated

▲ Old Picture of the Tanzhe Temple (Taken in 1920s)

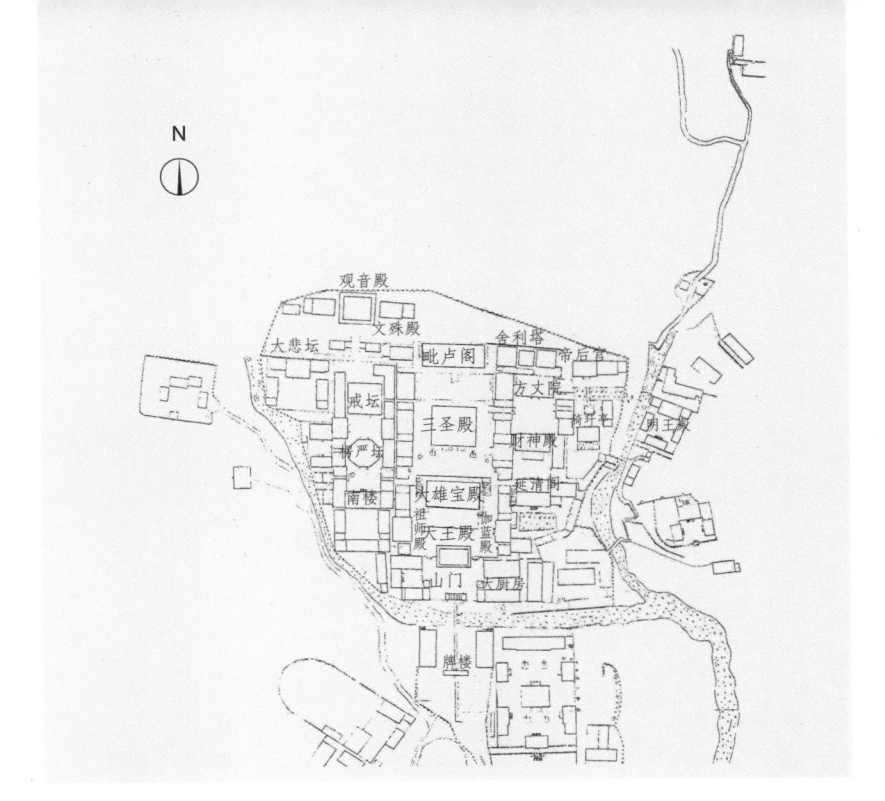

▲ General Plan Sketch of the Tanzhe Temple

in Empress Wu Zetian's reign. It was renamed the Grand Temple of Longevity under the reign of Emperor Huangtong in Jin and regained its name as the Jiafu Temple in 1457 under the reign of Emperor Tianshun in Ming. It was renamed the Xiuyun Temple in the 31st year (1692) under the reign of Emperor Kangxi in the Qing Dynasty. Although the name of the temple was changed many times, the most widespread name among the people was the Tanzhe Temple because "there is a Dragon pool behind the temple and cudriana trees on the mountain". Local people always call it the Tanzhe Temple (Tan refers to the pool while Zhe refers to cudrania trees).

The Tanzhe Temple was built on the south of Baozhu Peak and was surrounded by nine horseshoe-shaped gigantic surrounding peaks. The nine peaks from the east to the west are: Huilong Peak, Huju Peak, Pengri Peak, Zicui Peak, Jiyun Peak, Yingluo Peak, Jiayue Peak, Xiangwang Peak, and Lianhua Peak. The nine peaks are like nine huge dragons guarding the middle Baozhu Peak. The high peaks block the cold current sweeping from the northwest, making the area where the Tanzhe Temple is located form a warm and humid microclimate. So lush vegetation and a great variety of old trees and famous flowers are growing in the area surrounding the Tanzhe Temple and the environment there is rather beautiful.

▲ Panoramic View of the Tanzhe Temple

Buddhist Temples

▲ The Pailou (The Decorated Arch)

▲ Deep Valley in front of the Main Gate

The south-facing grand Tanzhe Temple covers a ground area of 2.5 hectares. With the surrounding areas, it covers an area of 11.2 hectares. With the forests and hills included, the total area of the Tanzhe Temple reaches more than 121 hectares. The layout of the Tanzhe Temple is arranged in three directions orderly: the central axis, the east and the west. A four-column and three-storied Pailou is located in the front. The beamless main gate is behind the bridge and is three bays wide with a one-eaved gable-and-hip roof covered with round gray tiles. Under the eave stands the stone stele with Emperor Kangxi's autograph "The Youyun Temple-Constructed by Imperial Command". Behind the main gate stands the Hall of the Heavenly Kings which is three bays wide with a one-eaved gable-and-hip roof covered with green glazed tiles. Under the roof hangs the board with Emperor Kangxi's autograph "The Hall of the Heavenly Kings". In the center of the hall is offered the Maitreya Buddha behind which stands the Veta Buddha. The statues of the Four Heavenly Kings are on the east and west sides. Situated in the center of the temple, the Mahavira Hall is the most important hall topping the construction rank. It is five bays wide with a double-eaved hip roof. The upper tier of the roof is covered with yellow glazed tiles and the lower tier is covered with yellow glazed tiles with green sheared edges. The main ridge of the hall is particularly tall. A pair of glazed Chi Wen was constructed on both sides of the main ridge in 1692. Imitating the shape of the legendary animal in the Yuan Dynasty, the two glazed Chi Wen are 2.9 meters tall. They are the tallest ones among all the decorations in the temples in Beijing. On both sides of the two legendary animals were engraved golden chains, named "golden ribbons" which were granted by Emperor Kangxi to the Tanzhe Temple. Sakyamuni Buddha is enshrined in the middle of the hall and the statues of the 18 Arhats are standing on both sides of Sakyamuni. Two flanking halls named the Garan Hall and the Hall of Patriarch are symmetrically situated on the east and west side of the hall. The Bell

▲ The Hall of the Heavenly Kings

▲ The Mahavira Hall

▲ The Bell Tower

▲ Chi Wen on the Roof of the Mahavira Hall and the Golden Chains

▲ The Vairochana Hall (The Pilu Hall)

and Drum Towers were built inside the hall. The Hall of the Three Holy Ones was once built behind the Mahavira Hall but its relics can only be seen now. Located on the highest spot of the Temple, the Vairochana Hall (Pilu Hall) is the last important hall and was once a place for storing Buddhist scriptures. It is seven bays wide with a gable-and-hip roof covered with round gray tiles.

▲ The Large Kitchen

▲ The Abbot Yard

On the eastern axis stand the Large Kitchen, the Yanqing Pavilion, the God of Wealth Hall, the Abbot Yard, the Emperor and Queen Hall and the Stupa Pagoda, etc. These buildings are mainly used for the monks' residence and for them to perform chanting. Rooms are also reserved for the Qing emperors to take a rest when making pilgrimages in the temple. Among all the temporary imperial halls, the Liubei Pavilion is very distinctive. It was also named the Yigan Pavilion, a place for the royal families to compose verses while drinking. Constructed on the site of the former Wuyi Hall, the pavilion has a quadrangle pavilion roof covered with glazed green tiles,

▲ The Yigan Pavilion in the Emperor and Empress Hall

▲ Elevation of the Altar of Rigor

under which hangs the board with Emperor Qianlong's autograph "Yigan Pavilion". The floor inside the pavilion was paved with white marble stones. On the floor a ten centimeter wide and ten centimeter deep sink was built. The sink takes a dragon's shape seen from south to north and a tiger's shape seen from north to south, so the pavilion is also called the Dragon Tiger Pavilion.

The building complex in the west of the Temple were laid out according to how they were used. On the south side is the altar-styled architectural complex which includes the Ordination Altar for holding the ceremony to ordain as monks, the Altar of Rigor and the Altar of Great Mercy for the senior monks to teach Buddhist scriptures, rooms for monks to copy Buddhist scriptures and Nan Hall for the senior monks. On the north side is the Buddha hall complex built near the hills

▲ Tap Supplying Water for the Yigan Pavilion in the Emperor and Empress Hall

▲ The Ordination Altar

which mainly includes the Goddess of Mercy Hall, the Manjusri Hall, the Hall of Patriarch and the Dragon King Hall, etc.

A hall used for the older monks to seek peace, happiness and longevity lies on the left front of the Tanzhe Temple. The east and west Goddess of Mercy Holes are on the east and west side behind the Tanzhe Temple. The Hall of Dharmapala and a complex of buildings lie in the east of the temple. Besides, in the front of the Tanzhe Temple are two monk tomb pagoda complexes where the senior monks from the Jin Dynasty to the Republic of China period were

▲ Emperor Tree

▲ Pagoda Forest

buried. The 75 tomb pagodas existent now are the precious materials for investigating Buddhism and the ancient architectures.

The Tanzhe Temple owns a great variety of precious affiliated cultural heritages, such as gold-plating decorations, stone tablets on which poems were inscribed in the Jin Dynasty, flesh Buddha in Qing Dynasty, huge bronze pots, mysterious stone fishes, the bricks used by princess Miaoyan for worshiping, etc.

Many famous ancient trees are another marvelous wonder of the Tanzhe Temple. Among them, 186 trees are on the national ancient tree preservation list. The most famous ancient trees include the two millennial ginkgos entitled "Emperor Tree" and the "Empress Tree" by Emperor Qianlong, the millennial salas and pines and the "Tree of Prosperity". There are more trees such as the Ming Erqiao magnolias, the "Gold Inlaid with Jade" and "Jade Inlaid with Gold" bamboos, black bamboos, and the ancient Jasminum floridum, etc. The ancient cultural heritages underwent numerous historical vicissitudes and possessed many marvelous legends, which greatly appeals to people and brings mysterious glories to the ancient temple.

Buddhist Temples

# The Yunju Temple

The Yunju Temple is located in Fangshan District of Beijing, 70 kilometers from downtown. The Yunju Temple is composed of the monastery, the stone-carved scriptures, the cave for storing Buddhist scriptures and the Tang and Liao pagoda complexes. It is well known for the preserved stone-carved Buddhist scriptures *The Tripitaka—The Stone-carved Buddhist Scriptures in Fangshan*. It was designated by the State Council as a major historical and cultural site under state protection in 1961.

The Yunju Temple was first built in the years of Emperor Daye's reign in the

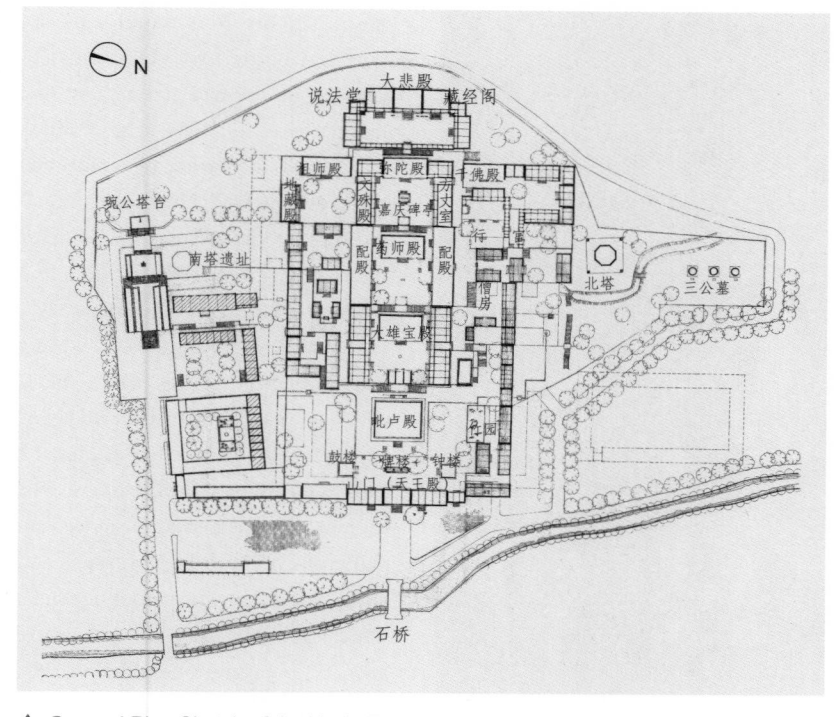

▲ General Plan Sketch of the Yunju Temple

▲ The Main Gate

Sui Dynasty. It had comprised five courtyards and six halls in the years of Emperor Sheng's reign in the Liao Dynasty. The temple went through renovations in the Jin, Yuan, Ming and Qing Dynasties. It was destroyed by the bombs of the Japanese Army in 1942. After the liberation of China, the government called on people to clean up the temple and set up a special organization to protect it. As a temple that preserves more stone-carved printing blocks for the Tripitaka than any other temple in the world, it was selected as "the world top tourist attraction in Beijing" in 1992. The pagodas in the Yunju Temple and the stone-carved scriptures hit the world cultural heritage tentative list in the same year.

The halls in the Yunju Temple were renovated with the financial support from the government and donations from the social organizations home and abroad. The original framework of the temple has been restored. It is composed of five courtyards including the Main Gate (the Hall of the Heavenly Kings), the Pilu Hall, the Mahavira Hall, the Hall of Bhaisa jyaguru, the Amitabha Hall and the Hall of Great Mercy.

▲ Stele Constructed under Emperor Jiaqing's Command in the Qing Dynasty

Two Liao pagodas stand opposite to each other in the south and north of the Yunju Temple. The south pagoda was also called the Pagoda . It doesn't exist now. Originally named the Stupa Pagoda, the 30 meters tall north pagoda was also called the Arhat Pagoda. The lower part of the pagoda is of a pavilion style with a hollow center. Stone steps can stretch out to its top. The upper part of the pagoda is of an inverted-alms-bowl shape and is covered by the rings around the steeple and by a roof. This building is of very unique characteristics among all the ancient pagodas extant in China. A multi-eaved Tang pagoda was built on each of the four sides of the north pagoda. These small pagodas were constructed during the period from the 2nd year (711) in Emperor Jingyun's reign to the 15th year (727) in Emperor Kaiyuan's reign. In the temple and on the surrounding mountains were built over ten brick pagodas and stone pagodas in the Tang, Liao and Ming Dynasties. Among them, the most famous two pagodas are the Wangong Pagoda that was the tomb pagoda for the founder of the stone-carved Tripitaka Jingwan and the Princess Jinxian Pagoda built for memorizing the Tang Princess Jinxian who contributed to the cause of scripture carving.

Although the Yunju Temple was destroyed by bombs in the past, the world famous stone-carved *Tripitaka* which was buried under the earth was perfectly preserved. The stone carved Tripitaka in the Yunju Temple was first carved during the year of Daye in the Sui Dynasty. The senior monk Jingwan carried on the unfulfilled wish of his master Huisi who practiced Buddhism on Mount Heng in the North Qi Dynasty. And he began to carve scriptures on the stone tablets on the Shijing Mountain (The Stone Scripture Mountain). After Jingwan died in the 13th Zhenguan year (639), his disciples Xuandao, Yimo, Huixian and Xuanfa etc. continuously took charge of the cause of scripture carving that went through the Tang, Liao, Jin, Yuan and Ming Dynasties and lasted about one thousand years. The printing blocks for the Tripitaka were stored in the nine caves on the Shijing Mountain (the Stone Scripture Mountain) and in the other underground halls for storing the stone-carved scriptures. Among them, the first and foremost place for storing the stone scriptures is the Fifth Cave called the Leiyin Cave (the Thunder Sound Cave). On the walls of the cave were embedded 146 blocks of stone carved scriptures initially engraved by Jingwan. In the center

▲ The North Pagoda

of the cave erect four octagonal stone columns. Each of the columns was engraved with Buddha reliefs totaling more than one thousand statues hence the columns were called the One-thousand Buddha Columns. Because the caves on the hill were completely filled with stone-carved scriptures, blocks of stone-carved scriptures were not carried up the hill any more but buried in the underground hall built on the southern side of the Yunju Temple. The Pagoda for Suppressing the Scriptures (namely the Southern Pagoda of the Yunju Temple) were built over the underground hall. With the support of the Chinese government and the Buddhists nationwide, the China Buddhism Association began to excavate the unique and important cultural heritage. It took them three years to complete the excavation and the inscription rubbing work. Afterwards, they gathered people to organize and study the scriptures and compiled the *Stone-carved Buddhist Scriptures in Fangshan* which is being published. In order to preserve the stone-carved scriptures in the Yunju Temple better, more than ten thousand blocks of the stone-caved scriptures were transferred to the underground hall. The stone-carved scriptures in the Yunju Temple contain more than one thousand engraved Buddhist classics, three thousand volumes of the scriptures and more than 14 thousand blocks of stone-carved scriptures. They are not only a great treasure for studying Buddhism, but also the material foundation for collating the extant wood-carved Buddhist classics because most of the manuscripts for the Tang stone carving were based on the imperial manuscripts contributed by the Tang Princess Jinxian and the Liao stone carving was based on the long-lost *Khitan Tripitaka*, the characters in these two versions are greatly different from the common versions of Buddhist classics. The stone-carved scriptures are imbued with great academic value for investigating the ancient society, exploring the development of the inscriptions on the ancient bronzes and stone tablets and the calligraphy.

The Yunju Temple also offers the Buddhist relics which were discovered under the Leiyin Cave in 1981. They were encased in five stone cases, silver cases and jade cases tier upon tier. The cover of the case was engraved with the date for setting the Buddhist relics and the case contains the jewelries for enshrining the Buddhist relics. The China Buddhism Association held a grand Dharma assembly in April 1987 to receive the Buddhists for worshipping and the international guests for sightseeing.

▲ The Tang Pagoda

▲ The Leiyin Cave

▲ Relief Sculpture in the Tang Pagoda

▲ The Stone Scripture Boards of the Liao and Jin Dynasties

Buddhist Temples

▲ The Stone Tablet Rubbing in the Leiyin Cave of the Tang Dynasty

# The Jietai Temple

The Jietai Temple is located on the Ma'an Mountain in Mentougou District of Beijing. Its official name is the Wanshou Chan Si (the Longevity Buddhist Temple). It is famous for the largest ancient ordination altar extant in China so it gets its name the Temple of the Ordination Platform or the Temple of the Ordination Altar. It was designated by the State Council as a major historical and cultural site under state protection in 1996.

The Jietai Temple was first built in the years of Kaihuang in the Sui Dynasty

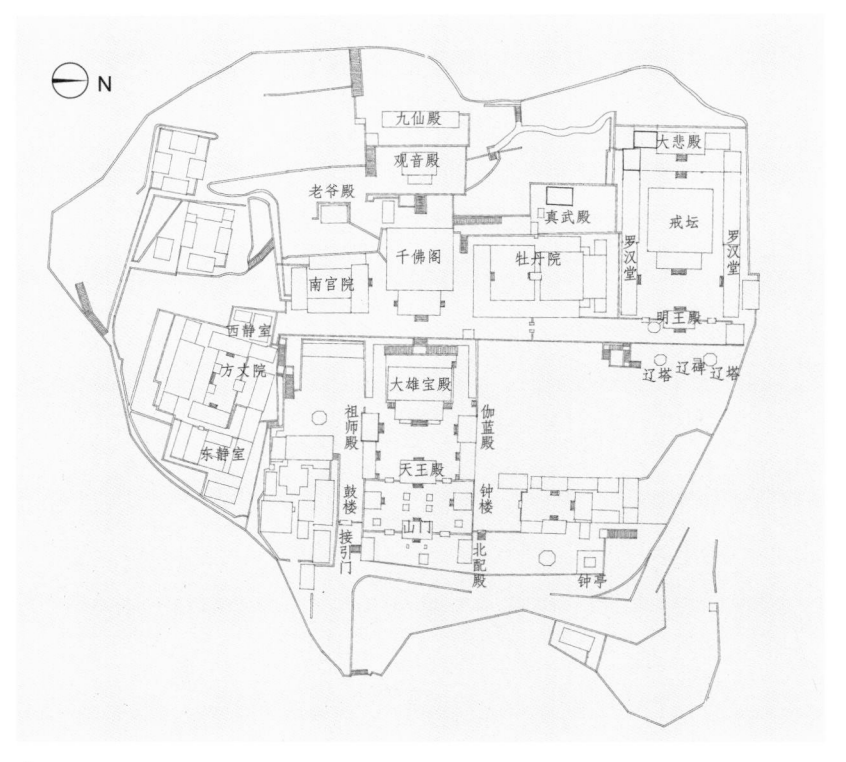

▲ General Plan Sketch of the Jietai Temple

and was named the Huiju Temple at that time. The Liao Emperor Daozong summoned the senior monk Fajun during the years of Xianyong and granted the *Mahayana Bodhisattva-pratimoksa* copied by himself to Fajun. This Bodhisattva-pratimoksa was regarded as a token of the orthodox Vinaya School, which promoted the Jietai Temple to be the most authoritative temple for imparting Buddhist doctrines and the holy land for the Vinaya School. The Jietai Temple was severely destroyed by the wars at the end of the Yuan Dynasty. It underwent several renovations conducted by the imperial family during the years of Xuande in the Ming Dynasty, which helped shape its present framework. Emperor Yingzong renamed it Wanshou Chan Si (The Longevity Buddhist Temple) and inscribed the name for the temple. He appointed Master Ruhuan to impart Buddhist doctrines here. The temple underwent seven-year large-scaled restorations during the years of Jiajing's reign and its size was further expanded. Emperor Kangxi and Qianlong went here for worshipping and offering incense several times during the years of their reign and ordered to protect and renovate the Jietai Temple. The temple was well preserved from the end of the Qing Dynasty to the Republic of China period. It was severely destroyed by the wars after the 1930s. After the foundation of the People's Republic of China, the government carried out several large-scale restorations on the Jietai Temple and opened it to the public.

▲ Old Picture of the Jietai Temple (Taken in 1920s)

▲ Panoramic View of the Jietai Temple

Constructed near the mountain, the east-facing Jietai Temple reflects the sun-facing construction style in the Liao Dynasty and covers an area of 4.3 hectares. The temple is composed of the southern and northern parts. All the halls were built on the gradually elevated terraces. Most of the halls extant were built in the Qing Dynasty or in modern times. In the southern part are situated the main structures of the temple. The main gate is three bays wide with a hip roof covered with round gray tiles. The stone tablet inscribed with the "Notes to the Ordination Altar in the Wanshou Temple" composed by Emperor Kangxi was set up in front of the main gate where also stands the south-facing Ksitigarbha Hall. The Hall of the Heavenly Kings behind the main gate is three bays wide with a hip roof covered with glazed green tiles. The Bell and Drum Towers were built on the two sides before the Hall of the Heavenly Kings. On the platform stands the Mahavira Hall that is five bays wide and three bays long with a flush gable roof covered with glazed green tiles. In the outer room hangs the board on which were written Emperor Qianlong's characters "Lian Jie Xiang Lin". The bronze statue of the Three Worlds Buddha molded in the Ming Dynasty was offered in the Mahavira Hall whose roof was

Buddhist Temples

decorated with three sunk panels engraved with flying dragons. On the south and north side of the Mahavira Hall are situated three side halls all with a flush gable roof covered with round tiles. The Garan Hall and the Hall of Patriarch are respectively standing on the north and the south side of the Mahavira Hall. Behind the Mahavira Hall erects the tall terrace where built the Ten-thousand Buddha Pavilion, the extant pavilion was rebuilt. Behind the Ten-thousand Buddha Pavilion comes a tall terrace where was constructed the Goddess of Mercy Hall that is three bays wide with a flush gable roof covered with round tiles. On the south side of the Goddess of Mercy Hall stand three rooms built for Prince Gong Yixin and used as his study rooms. The Jiuxian Hall was constructed behind the Goddess of Mercy Hall. On the south and north side of the Ten-thousand Buddha Pavilion stand two courtyards named the South Palace Yard and the North Palace Yard. They were a part of the temporary imperial palaces. The North Palace Yard was also called the Peony Yard whose architectural style perfectly combines the feature of the Beijing courtyards and that of gardens in southern China. It merges the primitive simplicity of the Beijing courtyards with the elegance of the southern China gardens. Emperor Qianlong together with the consorts once stayed here for escaping the summer heat and Prince Gong Yixin once lived here for a long time. On the south side of the South Palace Yard were built two small courtyards the Upper Courtyard and the Lower Courtyard which were used for the monks' residence. On the south side of the Mahavira Hall stands an abbot courtyard including two yards. On the east and west side of the Mahavira Hall are respectively standing the east and west ashrams.

▲ The Main Gate

▲ Stone Rails and Pine Trees before the Main Gate

Buddhist Temples

▲ The Mahavira Hall

▲ The Goddess of Mercy Hall

▲ Ornamental Gate in the Peony Courtyard

027

▲ The Hall of the Ordination Altar

The Ordination Altar Courtyard stands in the northern part of the Jietai Temple. The Hall of Dharmapala is three bays wide with a gable-and-hip roof serving as the main gate of the Ordination Altar Courtyard. In the front of the Hall of Dharmapala stand two octagonal stone Dhanari columns. The Ordination Altar Courtyard contains the Ordination Altar Hall that was first built in the 5th year (1069) of the Liao Xianyong's reign and underwent renovations in the Jin, Yuan, Ming and Qing Dynasties. With a square shape, the Ordination Altar Hall is five bays wide and five bays deep with a double-eaved quadrilateral roof covered with round tiles. On the terrace in the center of the roof is seated a bronze gold-glided block laid out in a throne pagoda style. Located in the center of the hall, the Ordination Altar is in a blue stone masonry structure and in a square layout. The whole altar is 3.25 meters tall and divided into three stories: the lower story is 1.4 meters tall and each of its four sides is 11.3 meters long; the middle story is 0.95 meters tall and 9.6 meters long, the upper story is 0.9 meters tall and 8.1 meters long. The terrace of each story has a high-molding throne base and the waist of the terrace was engraved with a niche in which the statue of the Commandments of God is offered. As the largest and most elegant Ordination Altar, it has 113 niches and 113 statues of the Commandments of God. The Ordination Altar Hall once contained some agilawood-made chairs engraved with designs serving as the seats for the Three Masters and Seven Certifiers when they were imparting Buddhism but they don't exist

▶ The Ordination Altar

Buddhist Temples

▲ The Liao Pagoda

any longer. Ten rosewood chairs and tables were put in the hall. The ceiling of the hall was decorated with eight exquisitely carved sunk panels that were the magnificent artworks among all the extant sunk panels. On each of the two sides of the hall stand 18 side halls which served as the Five-hundred Arhat Hall. 500 clay-sculptured and painted Arhats were placed in this hall but they are non-existent now. Behind the Ordination Altar Hall erects the Hall of Great Mercy that is five bays wide with a gable-and-hip roof covered with round tiles.

On the east side of the Jietai Courtyard stand the Jietai Pagoda complex. Under the terrace of the Jietai Courtyard stand two paralleled south and north brick pagodas. Built in the 1st year (1075) under the Liao Dakang's reign and renovated in the 13th year (1448) under the Ming Emperor Zhengtong's reign, the North Pagoda is the tomb pagoda for Fajun. It was a seven-story multi-eaved octagonal pagoda. To the south of the North Pogada erects a stele inscribed with the characters "Fa Jun Da Shi Yi Xing Bei" (Master Fajun's Trace Stele) written by Wang Ding in the 7th year under the Liao Emperor Da'an's reign. As the mantle and alms bowl pagoda for Fajun, the octagonal multi-eaved South Pagoda has five stories.

# The Shifang Pujue Temple

The Shifang Pujue Temple is located inside the Beijing Botanical Garden on the Wofosi Road in Haidian District, Beijing, at the south foot of the Jubaoshan Mountain. Because of the largest bronze reclining Buddha in the temple in China, it is also called the Wofo Temple (the Temple of the Reclining Buddha). The temple was designated by the State Council as the major historical and cultural site under state protection in 2001.

Built during the Zhenguan period of the Tang Dynasty, the Shifang Pujue Temple was originally named the Doushuai Temple. It was expanded in the Yuan Dynasty and renamed the Dazhaoxiao Temple. The Yuan Emperor spent a huge amount of money renovating the temple and casting the statues of Buddha in the 1st year (1321) during the Zhizhi period. The restoration work cost five million taels of silver. Thousands of craftsmen worked for 12 years. In order to cast the

▲ Avenue in front of the Temple

▲ The Glazed Pailou

reclining Buddha, "a huge amount of money was invested and seven thousand soldiers were called on to support the construction" (according to the historical records). It was renamed Shou'an Chan Lin (The Temple of the Peaceful Longevity) after its renovation in the 8th year (1443) during the Zhengtong period in the Ming Dynasty. It was renamed the Yong'an Temple (The Temple of Permanent Peace) in the 18th year (1482) during the Chenghua period in the Ming Dynasty. The temple was renovated in the 12th year (1734) during the Yongzheng period in the Qing Dynasty and was renamed the Shifang Pujue Temple (Shifang in Buddhism means the ten directions and Pujue means universal awakening), which is continued to be used today. The structures extant were all built in the Qing Dynasty.

In the front of the temple stands a three-storied and four-columned wooden archway whose roof was covered with gray tiles. The tablet on the archway was inscribed with the characters "Zhi Guang Chong Ming" (enlightenment of wisdom and perspicacious understanding). After passing the archway, visitors can see a 100 meters long avenue. Behind the avenue stands a huge glazed archway that is seven-storied and four-columned with a roof covered with green glazed tiles and carrying with sheared yellow edges. The tablet on the archway was inscribed with the characters "Tong Can Mi Zang" (consult together the mystery of Buddhism)and "Ju Zu Jing Yan". Behind the archway stands a semi-cylindrical pond served as a pond to memorize the great achievements of the Buddhists. A bridge is crossing over the pond. After crossing the bridge, visitors can see the main gate of the temple where the Bell and Drum Towers are standing on the left and right. The structures laid out on the central axis behind the main gate include the Hall of the Heavenly Kings, the Hall of the Buddha of the Past, the Present and the Future, the Hall of Reclining Buddha and the Hall for Storing Scriptures. The side halls such as the Dharma Hall and the Prince Siddartha Hall are standing on the two sides of the main halls. Flanking doors were opened in the main gate walls. Galleries were built outside the flanking doors. The two flanking galleries, form a lengthwise cloister, which shows the temple's layout at the early stage. Such a layout is rarely seen in Beijing.

The structures in the east and west are standing on the two sides of the central axis. The east courtyard includes the monks' residence place such as the Abstinence Hall, the Meditation Hall, the Jiyue Pavilion and the Courtyard of Patriarch, etc. The structures in the west are composed of five courtyards that originally served as the temporary imperial palace for the emperor to escape the summer heat and deal with the government affairs. The temporary imperial palace also includes the artificial hills, stone bridges and ponds etc.

The middle part of the Temple of Reclining Buddha forms a courtyard, several

▲ Carvings on the Glazed Pailou

▲ Stone Board on the Glazed Pailou

▲ The Stone Bridge and Main Gate

▲ The Hall of the Buddha of the Past, the Present and the Future

▲ The Hall of Reclining Buddha

Buddhist Temples

035

▲ The Reclining Buddha

▲ The Open Hall of the Imperial Temporary Dwelling Palace

▲ The Square Pavilion of the Imperial Temporary Dwelling Palace

courtyards in the eastern and western parts, the southern and northern lanes. Such a layout reflects the remaining cloister building standard passed down from the Tang and Song Dynasties. It shows where were the courtyards and corridors often built in the two dynasties. The structures in the Wofo Temple have a great value for studying the evolution of the ancient temples.

Buddhist Temples

037

# The Fayuan Temple

Situated at No. 7 Fayuansi Front Street in Xuanwu District, Beijing, the Fayuan Temple is one of the city's most renowned Buddhist temples. It was designated by the State Council as the major historical and cultural site under state protection in 2001.

Located in the southeast of Youzhou County, the Fayuan Temple was first built in the Tang Dynasty. In the 19th year (645) of Zhenguan, the second Emperor Li Shimin, wanted to build a temple there for mourning the soldiers died in the northern expedition to Liaodong (the southeast part of Liaoning Province now), but his wish wasn't fulfilled. The temple was not completed until the 1st year (696) of Tongtian, the reign of Empress Wu Zetian. It was named the Minzhong Temple (The characters Ming and Zhong mean Loyalty). The temple had always been a famous building in north China until the Liao and Jin Dynasties ever since it was built. It was destroyed in the Yuan and Ming Dynasties. The eunuchs donated to rebuild the temple in the 3rd year of Zhengtong's reign (1438) and renamed it Chongfu Temple. The size of the present temple was formed at that time. Compared to the temple in the Tang and Liao Dynasties, half of its area was reduced. Most of the present architectures were built in the Qing Dynasty.

The main gate of the south-facing Fayuan Temple has three sub-gates with a gable-and-hip roof and the middle gate functions as the main entrance. A pair of

▲ The Screen Wall

▲ The Main Gate

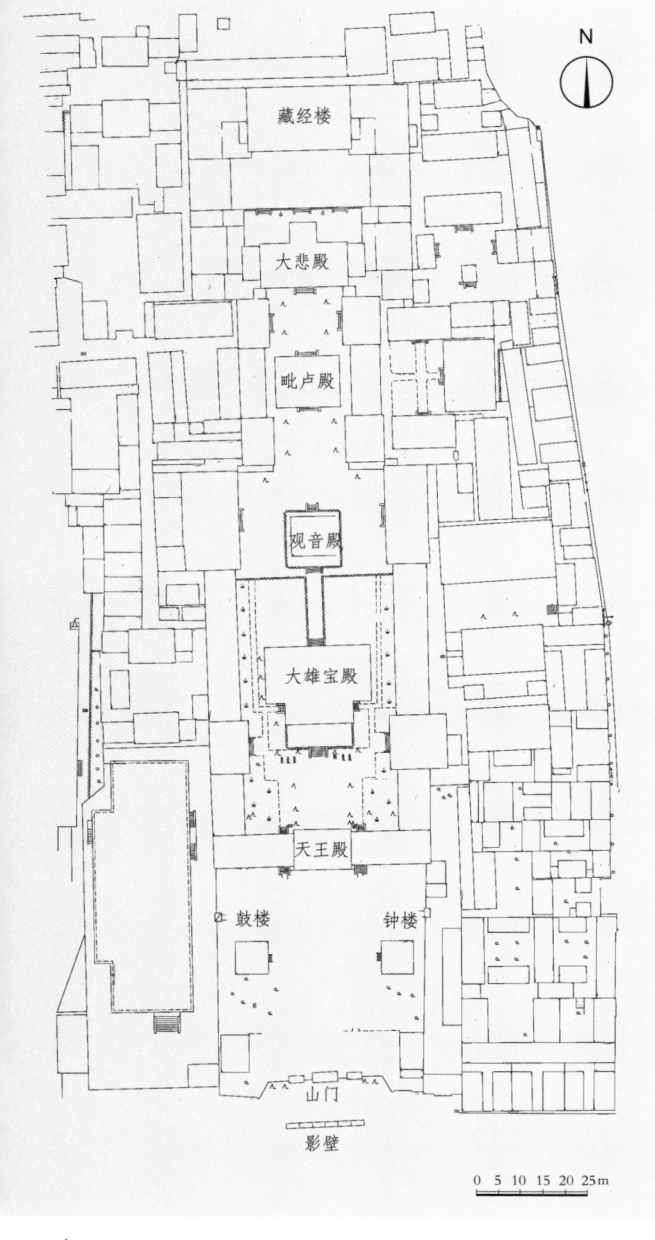

N

藏经楼

大悲殿

毗卢殿

观音殿

大雄宝殿

天王殿

鼓楼　　钟楼

山门

影壁

0 5 10 15 20 25m

▲ General Layout of the Fayuan Temple

stone lions stand in the front of the gate. The two side doors both have an overhanging gable roof covered with round tiles and supported by a main ridge. Splay screen walls were built on the two sides of the Main Gate. A brick one-glaph-shaped screen wall is standing opposite to the main gate. Behind the entrance of the Main Gate are located the Bell and Drum Towers on the east and west side of the temple. In the center of the temple stands the Hall of the Heavenly Kings which is three bays wide and seven purlins deep. With sandstone-windowed front eaves, it has a flush gable roof covered with round tiles and suspended by a main ridge. Galleries with side doors on the left and right sides of the Hall of the Heavenly Kings extend northward and reach the last building, the Canon House, which forms a closed hall surrounded by side halls and rooms with a width of 50 meters from east to west and a length of 180 meters from south to north. In the center of the Hall erect five buildings including the Mahavira Hall, the Goddess of Mercy Hall, the Pilu Hall (the Ordination Altar for holding the ceremony to ordain as monks), the Hall of Great Mercy and the Pavilion for Storing Buddhist Scriptures.

The main hall in the five bays is

Buddhist Temples

▲ The Mahavira Hall

▲ Interior View of the Mahavira Hall

five bays wide and seven purlins deep Mahavira Hall that has a gable-and-hip roof covered with round tiles. There are three verandas and a brick masonry platform in the front of the hall. On the two sides of the platform erect six steles. In the east and west were built three side halls with flush gable roofs supported by a main ridge and covered with round tiles. A gallery stands in the front of each hall. The main hall in the second courtyard is the Goddess of Mercy Hall that is three bays wide and deep with a square layout. This hall was based on a high base which was said to be the base for the main hall of the Minzhong Temple. Brick guardrails are surrounding the base. With a flush gable roof supported by a main ridge and covered with round tiles, the Goddess of Mercy Hall has five side halls on its east and west, a gallery in the front of it. The main hall in the third courtyard is three-bay wide Pilu Hall with a gallery in the front of it and a flush gable roof supported by a main ridge and covered with round tiles. On each of the east and west side were laid out three side halls all with front galleries and a flush gable roof supported by a main ridge and covered with round tiles. The circuit rooms on the two sides of the Pilu Hall can access to the small east and west yards. The main hall in the fourth courtyard is the Hall of Great Mercy, five bays wide and seven purlins deep with a flush gable roof supported by a main ridge and covered with round tiles. A veranda with a round

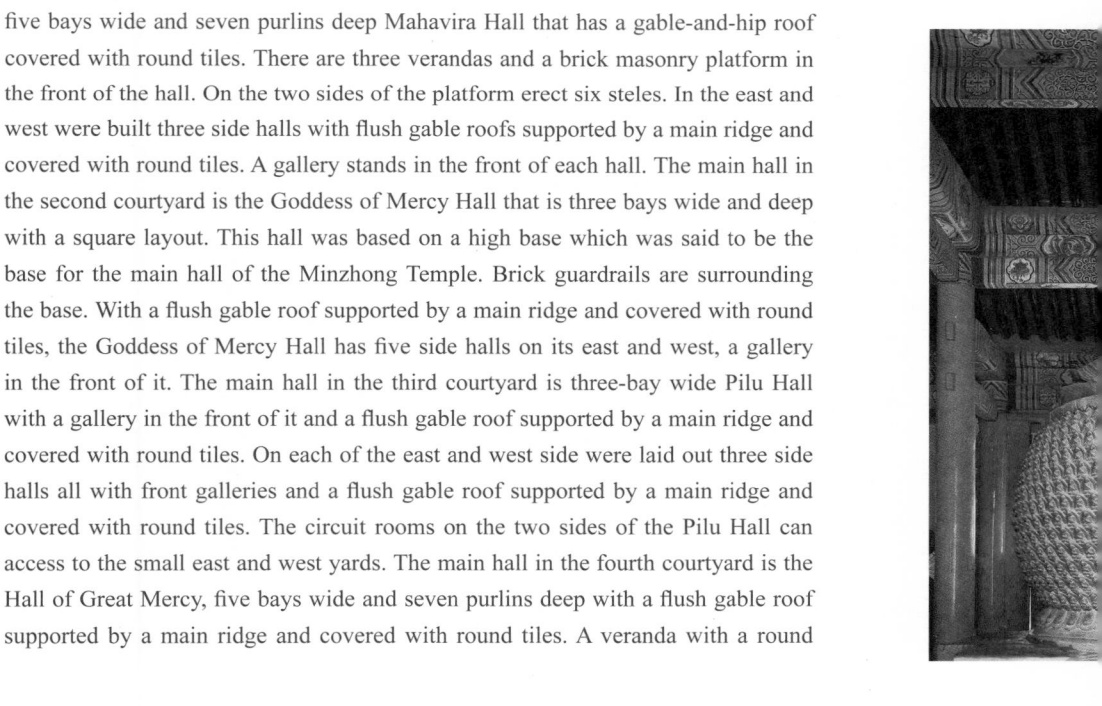

▲ The Goddess of Mercy Hall

▲ Interior View of
the Pilu Hall

▲ The Pavilion for Storing Buddhist Scriptures

Buddhist Temples

041

▲ Section of Shi Siming's Stele of the Tang Dynasty

▲ Base of the Column of the Tang Dynasty

ridge roof and suspended ridge follows the main hall. A flanking hall connects each of the two sides of the main hall with three side halls standing in its east and west. The fifth courtyard comes to the two-story Hall for Storing Buddhist Scriptures that is five bays wide with a gable-and-valley roof. On the two sides at the back of the Hall for Storing Buddhist Scriptures were built corner buildings connecting the side halls.

There are three buildings on the left side of the temple. The one in the north serves as an abstinence hall that is three bays wide with an overhanging gable roof covered with round tiles. The other two serve as the monks' residence.

Two towers were built in the east and west of the Minzhong Temple in the Tang Dynasty. The eastern tower was built under command of An Lushan and the western tower was constructed under command of Shi Siming. The towers had been ruined long ago and what was left in the temple is the preserved inscription board by Shi Siming when the tower was built, showing the long history of the temple. The construction of the tower can be testified from the recorded details in the inscription. Besides, under the golden column in front of the gate of the Mahavira Hall are two round lotus-blossom column bases which were exquisitely carved and might be left behind by the Liao Dynasty or the Tang Dynasty.

The China Buddhist Literature and Heritage Museum was set up in the Fayuan Temple in 1979 and it has become an important organization to collect, investigate and exhibit the Buddhist heritages, books, doctrines and the history of Buddhism.

# The Dajue Temple

The Dajue Temple is located at the south foot of the Tangtai Mountain, Haidian District and is one of the most well-known temples in Beijing. It was designated by the State Council as the major historical and cultural site under state protection in 2006.

The Dajue Temple was first built in the 4th year (1068) in Xianyong's reign in the Liao dynasty. Deng Conggui, a Buddhist in Nanyang, donated a huge amount of money to construct this temple. *Tripitaka Sutra* was inscribed. The Dajue Temple was originally known as the Clean Water Temple and became the temporary imperial palace in the Jin Dynasty. It was one of the eight important temples during the emperor Zhangzong's period in the Jin Dynasty. The temple was renovated and renamed the Dajue Temple in the 3rd year (1428) under Xuande's reign in the Ming Dynasty. During the late Ming and early Qing Dynasties, this temple was completely destroyed. It received large scale reconstruction in the 59th year (1720) in Kangxi's reign in the Qing Dynasty. The reconstruction was supported by the prince Yong who later became the successive emperor Yongzheng. The Siyi Hall and the Lingyao Pavilion were built up then. It was renovated extensively in the 12th year (1747) in Emperor Qianlong's reign. The White Dagoba was built under the imperial command.

The east-facing Dajue Temple reflects Liao people's worship of the Sun. The temple was constructed in accordance to the mountain's landscapes. Temples were built in the central axis. Temporary imperial palaces were built in the south part while in the north part were monks' dormitories.

With a gable-and-hip roof, the three-bay wide main gate serves as the entrance. In the middle of the main gate is an arch. Six Chinese characters meaning "the Dajue Temple constructed by imperial command" were inscribed on the board which hung above the arch. In the first set of yard is located the Hall of the Heavenly Kings where Maitreya is enshrined. The hall is also known as the Maitreya Hall. Inside the hall Four Heavenly Kings are enshrined on both sides. Go through the Hall of the Heavenly Kings visitors arrive at the Mahavira Hall which

▲ The Main Gate

▲ The Free Life Pond near the Stone Bridge

is five bays wide and five bays deep with a gable-and-hip roof covered with glazed green tiles and a bracket set was installed under its eave. Behind the hall were three small buildings. In front of the hall is a base on which the Buddhas of the Past, Present and Future are enshrined. They were removed here from the Zhihua Temple in 1972. On each side of the Hall are three side halls. The Hall of Amitayus which is five bays wide, five bays deep with a gable-and-hip roof covered with glazed

▲ The Mahavira Hall

▲ The Hall of the Amitayus Buddha

▲ The Suspended Sculptures in the Hall of the Amitayus Buddha

Buddhist Temples

045

green tiles is the third hall in the central axis. A brick pathway connects the hall to the Mahavira Hall. In the hall are enshrined a Buddha and two bodhisattvas. On the wall behind the Buddha statue is classic clay works on which the bodhisattva of Compassion is seated among mountains and sea. In front of the hall are two ginkgo trees, one of which is over a thousand years old. It has got reputation of "Ginkgo Tree King". On each side of the hall are ten side halls. The Hall of Great Mercy, also known as the Altar of Great Mercy is the fourth hall. The two-story building is five bays wide and six bays deep with a flush gable roof. North to the hall is a stele which was erected in the 4th year under the Emperor Xianyong's reign in the Liao Dynasty (1068). On the stele was inscribed history of the temple and history of inscription of Tripitaka Sutra. It is the precious relics in the temple. On the slope behind the hall was built the White Pagoda.

The two-story timber structured Dragon King Hall is located at the end of the temple's central axis. There is a pond surrounded by octagonal white marble stone

▲ The Hall of Great Mercy

▶ The White Pagoda

▶ The Hall of the Dragon King and the Pond

Buddhist Temples

047

guardrails and filled with clear spring water sloping down the steps. In front of the hall stand several cypresses.

The first building in the south road is the Ordination Hall. The second set of courtyard is known as the Siyi Hall, also known as South Yard of Magnolia denudata. Built in the years in the Prince Yinzhen's reign, it was named after the Emperor Yongzheng's study. In the courtyard stands an over 300-year Magnolia denudata, known as "Magnolia denudata King" . The Qiyun Hall is situated in the third courtyard. The Lingyao Pavilion is situated at the west end of the road.

The monks live in the north courtyard on the north road. The first courtyard is the yard for the abbot. The south facing yard consists of a front yard and a rear yard. In the front yard there is a screen wall and two old Aesculus chinensis trees. In the rear yard stand a pine and a cypress. The second set of courtyard is the North Yard of Magnolia denudata. In front of the main building is a pond where springs from the Dragon Pond converge. A ginkgo tree is over ten meters, surrounded with nine small ginkgo trees, which looks like a small patch of wood.

▲ The Siyi Hall

▲ The Lingyao Pavilion

# The Badachu Park on the Western Hills

The Badachu Park (The Eight Great Sites Park on the Western Hills) is located at Sipingtai Village in Shijingshan District. Known as "The Eight Great Temples" in the past, the park has eight temples scattered on the Cuiwei Hill, the Lushi Hill, and the Pingpo Hill. The Badachu Park includes the famous ancient temple complex in western Beijing. Monks started to live there as early as in the Sui Dynasty. Most of the extant temples were constructed in the Ming and Qing Dynasties. They have become important scenic spots and Buddhist activity venues. It was designated by Beijing Municipal Government as the major historical and cultural site to be protected in 1957.

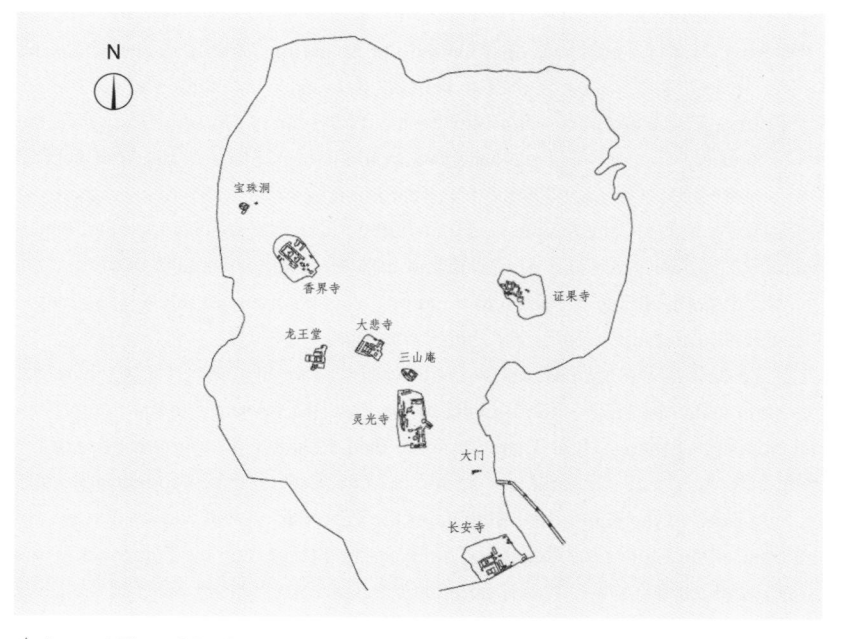

▲ Layout Plan of the Badachu Park on the Western Hills

The Eight Great Sites are distinguished for "three hills, eight temples and twelve sceneries". The west, north and east sides of the Badachu Park are surrounded by mountains and its south side faces a plain. With steep cliffs and gullies, luxuriant forests and clear water in it, the Badachu Park has a beautiful, quiet and pleasant environment. The temples are among the green pines and verdant cypresses. The eight temples laid out orderly are: the Chang'an Temple (the Temple of Eternal Peace), the Lingguang Temple (the Temple of Divine Light), the Sanshan Nunnery (the Three-hill Nunnery), the Dabei Temple (the Temple of Great Mercy), the Dragon Spring Nunnery, the Xiangjie Temple (the Temple of the Fragrant World), the Pearl Cave (the Cave of Precious Pearl), and the Zhengguo Temple. The first two temples are located at the foot of the hill, the third temple is located at the piedmont, the next three temples are located at the mountainside, the seventh temple is located at the peak and the eighth temple is located at the hill opposite to the hill where the seventh temple erects. Besides the eight religious temples, some other historic scenic spots that can be seen along the park are the Shuixin Pavilion, the Crane-releasing Pavilion, the Copper Buddha Cave and the Spyglass Pavilion.

The first Great Site of the Badachu Park is the Chang'an Temple (the Temple of Eternal Peace) which was also named the Shanying Temple or the Shanying Chang'an Temple. Located on the Cuiwei Hill, it was also called the Cuiwei Temple in the past. The temple was first built in the 17th year in Hongzhi's reign in the Ming Dynasty. Later, two large-scaled renovations took place in the 16th year in Emperor Shunzhi's reign and in the 10th year in Emperor Kangxi's reign in the Qing Dynasty. As a result, the temple was transformed into a large-scaled temple, which contained various halls and pavilions and had all kinds of Buddha and religious utensils. It gradually developed into a famous temple among all the temples on the Western Hills during the Ming and Qing Dynasties.

The east-facing temple has two courtyards. From the east to west, there are the Main Gate, the Hall of the Buddha of the Past, the Present and the Future, and the Goddess of Mercy Hall. There are more than 30 side halls on both sides of the main hall. The two courtyards can be walked into through two moon-shaped gates on both sides of the Hall of the Buddha of the Past, the Present and the Future. In the front of the Hall of the Buddha of the Past, the Present and the Future stand two wondrous luxuriant pines. It was said that the pines are the lead pines in the Ming Dynasty and have a history of more than 700 years. In addition, there are also many precious trees and flowers such as magnolias, Crape Myrtle trees, Papaya trees, Chinese redbuds etc. The statues in the Chang'an Temple are the most exquisite ones among all the temples. The craftsmanship of the statues in this temple is well shown in the *Record of the Royal Scenic Spots*: "The Buddha in the halls in the Shanying Temple are seated solemnly, with a sergeant appearance."

▲ The Main Gate

▲ The Hall of the Buddha of the Past, the Present and the Future

Buddhist Temples

Located at the east piedmont of the Cuiwei Hill, the Lingguang Temple, also known as the Temple of Divine Light is the second Great Site and also one of the most important temples among these eight temples. Constructed in the Tang Dynasty, it was named the Dragon Spring Temple. Renovated in the 2nd year (1162) in Emperor Dading's reign, it was renamed the Jueshan Temple. It was renovated again in the 14th year (1478) in Emperor Chenghua's reign in the Ming Dynasty and named the Lingguang Temple. The temple was destroyed by the Eight-Power Allied Forces in 1900 and most of the extant buildings were rebuilt later.

The south-facing Lingguang Temple had a lengthwise layout. The Great Mercy Hall and the Goldfish Hall are located in the south. The Abbot Hall, the Pagoda Complex, the Lay People Hall and the Mahavira Hall are situated in the north of the temple. There was once a Zhaoxian Pagoda (only its base can be seen now) in Lingguang Temple in which the monks found out a stone envelop with an agilawood-made box in it when they were cleaning the ruins of the temple. In the box a tooth from Sakyamuni was found. The tooth had been enshrined for 830 years from when it was brought into the pagoda in 1071 to 1900 when it was refound. In order to consecrate the divine tooth, the Buddhist Association of China built a

▲ The Base of the Zhaoxian Pagoda

thirteen-story octangular "Divine Tooth Stupa Pagoda" with multi eaves and a height of 51 meters in 1959. The Goldfish Pond was one of the rare relics preserved. The spring flows into the pond through the mouth of the stone-carved dragon. A delicate stone bridge crosses over the pond and a water-surrounded pavilion was built in the center of the pond. The tomb of Cuiwei who was the princess of the Yuan Dynasty, lies to the left of the Goldfish Pond, the Goddess of Mercy Cave and the stone fall lie to the north while the Taoguang Nunnery is standing at the back.

▲ The Tooth Relic of Buddha Pagoda

▲ The Thousand-hand Goddess of Mercy Hall

▲ The Main Hall

The third Great Site of the Badachu Park is the Sanshan Nunnery (the Three-hill Nunnery). It was first built in the 3rd year (1151) in Emperor Tiande's reign in the Jin Dynasty and was commonly known as the Majia Nunnery. It was named the Three-hill Nunnery as it was situated just among the Cuiwei Hill, the Pingpo hill and the Lushi Hill. The nunnery was renovated during the period of Emperor Qianlong in the Qing Dynasty.

The Three-hill Nunnery facing northeast is a delicately laid-out and well-designed courtyard. The three-bay hall in the main gate has a corner gate on both the left and right sides. The main hall has five bays and two side rooms. The two side halls are located on the two sides of the main hall with their doors opposite to each other. In the front of the main hall, there is a rectangle marble stone with water-flowing and cloud-floating designs; hence it is also called "the Water and Cloud Stone". At the back of the east wing-room is an open hall built on the tall terrace. A board inscribed with the characters "Cui Wei Ru Hua" (the Cuiwei Temple in the Painting) was hanged on the hall.

▲ The Water and Cloud Stone

The fourth Great Site of the Badachu Park is the Dabei Temple, which is located at the mountainside of the Pingpo Hill. It used to be known as the Yinji Temple. The temple was first constructed in the Jin and Liao Dynasties. The Dabei Hall was added in the 29th year (1550) in Emperor Jiajing's reign in the Ming Dynasty. It was renamed the Dabei Temple in the 51st year (1712) in Emperor Kangxi's reign in the Qing Dynasty and underwent a renovation in the 16th year (1795) in Emperor Qianlong's reign.

The east-facing Dabei Temple has three halls both in the front and at the back. It was constructed according to the topography of the mountain. The Main Gate is three bays wide. The lintel of the arched doorway was inscribed by Kangxi, on which were written "Chi Jian Da Bei Si" (Construction of the Dabei Temple by Imperial Command).Behind the Dabei Temple stands the Mahavira Hall where Buddha Sakyamuni and his two disciples Ananda and Kasyapa are seated in the center of the hall. The Eighteen Arhats are seated in the two sides. The Dabei Hall is at the back and was newly built in the 29th year (1550) under the reign of Qing Emperor Jiajing. Avalokitesvara Bodhisattva is seated in this hall. The newly-renovated Hall of the Medicine Guru Buddha is the last hall in the Dabei Temple. The pagoda complex of the Dabei Temple lies to the north of the temple and a brick pagoda in an inverted-alms-bowl style is still extant.

▲ The Main Gate

▲ Interior View of the Dabei Hall

▲ The Dabei Hall

To the northwest of the Dabei Temple lies the fifth Great Site named the Longwangtang Hall (the Dragon King Hall). In the late Ming and early Qing Dynasties, two temples were built there. One was the Huiyun Nunnery and the other was the Dragon King Hall. The previous one was built in the years in the reign of the Ming Emperor Hongxi and the latter was constructed in the 2nd year in the reign of the Qing Emperor Shunzhi. The nunnery and the Longwangtang Hall were combined into one nunnery during the period of the Qing Emperor Daoguang.

There are five courtyards in this nunnery, which are divided into the three levels: the upper, the middle and the lower level. Besides the Longwangtang Hall, the main hall of the nunnery, more structures include the Woyou Pavilion, the Waterside Pavilion, the Miaoxiang Nunnery and the Huazu Hall. The Longwangtang Hall is famous for its ancient cypresses and spring. What fascinates people is its two cypresses planted at the gate of the hall and called "Tree Flags" by the people. The east-facing temple lie two courtyards which house the clay statues of the Dragon King, the Thunder God and the Lightening Goddess. The spring water flows out from the back wall of the temple and pours into the pond through the stone-carving dragon at the courtyard on the first level. On the left side of the pond is the "Waterside Pavilion" which provides a good site for the tourists to take a rest. The back hall, the Woyou Pavilion, provides the best place for appreciating the red autumn leaves in deep autumn.

▲ The Main Gate

▲ Waterside Pavilion

▲ The Main Gate

▲ The Yuantong Hall

The sixth Great Site, the Xiangjie Temple (The Temple of the Fragrant World), lies in the northwest of the Longwangtang Hall. It is also known as the Pingpo Temple because it is located at the Pingpo Hill. The Xiangjie Temple is the largest temple among all the eight Great Sites of the Badachu Park. Built during the years of the Tang Emperor Qianyuan, it was named the Pingpo Dajue Temple. Later it was renamed the Dayuantong Temple in the Ming Dynasty and the Shenggan Temple by the Qing Emperor Kangxi. The temple was renovated in the 13th year (1748) in the reign of the Qing Emperor Qianlong and renamed the Xiangjie Temple.

The temple has three parts and houses five courtyards. The three-bay Main Gate with an arched marble stone doorway is located in the front of the first part. Under the eaves was inlayed with a white marble lintel on which were written "Chi Jian Xiang Jie Si" (construction of the Xiangjie Temple by imperial command) by Emperor Qianlong. Behind the Xiangjie Temple stand the Mahayana Gate(destroyed) and the Bell and Drum Towers. Next comes the Hall of the Heavenly Kings whose arched doorways and sill wall windows were made of delicately carved white marbles. The hall houses the potbellied Maitreya with the clay statues decorated with colored paintings of the Four Heavenly Kings on its both sides. Behind the Hall of the Heavenly Kings stands the Yuantong Hall. Two tall and delicately carved steles stand in front of the Yuantong Hall. The stele in the east has a turtle-shaped base and dragon-shaped top. The south-facing side of the stele was engraved with the inscription "Construction of the Shenggan Temple by Imperial Command". The north-facing side of the stele was engraved with the inscription "Construction of the Xiangjie Temple by Imperial Command". The stele was renovated in the 14th year in Emperor Qianlong's reign and was set up after the completion of the Shenggan Temple. In the west stands a huge bluestone stele whose base was delicately carved with spotted deer and sea horse designs. The characters "Da Bei Pu Sa Zi Zhuan Zhen Xiang" (the True Image of the Buddha of Great Mercy) were engraved on the south-facing side of the stele and the two characters "Jing Fo" (Worship Buddha) were engraved on the north-facing side. The two solemn characters are as big as a round arch and were inscribed by Emperor Kangxi. The Mahavira Hall that is five bays wide. Its doors and windows are carved with exquisite designs. The gold-lacquered Buddha of the Past, the Present and the Future and the statues of the Eighteen Arhats are enshrined in the hall. The last hall is the court for storing Buddhist scriptures. All the buildings in the court have two stories. The five-bay court has six side rooms on both sides. On the east side stands the "Imperial Palace" used by Emperor Qianlong as a summer resort. A courtyard with three rooms was built on the west side.

▲ The Shenggan Stele Constructed under Emperor Kangxi's Command in the Qing Dynasty

▲ The Pavilion for Storing Buddhist Scriptures

Buddhist Temples

▲ Ornamental Gate in the Imperial Temporary Dwelling Palace

▲ The Main Hall in the Imperial Temporary Dwelling Palace

▲ The Theatre Stage in the Imperial Temporary Dwelling Palace

▲ The Pailou

Located at the peak of the Pingpo Hill, the seventh Great Site, Baozhu Cave (Pearl Cave or sometimes translated as The Cave of Precious Pearl) was constructed in the 46th year (1781) in the reign of the Qing Emperor Qianlong. There is a wooden Pailou in the front of the temple and a board hangs above it with the characters "Jian Gu Lin" (Firm Forest) and "Huan Xi Di" (Joyful Land) written by Emperor Qianlong. There is a huge natural stone in the front of the Pailou. A poem entitled *Baozhu Cave Poem* in running script can been seen on the stone inscribed with the mark of Emperor Qianlong's Great Seal. The Baozhu Cave has two halls.

The Goddess of Mercy Hall is the main hall with two side halls on both sides. A four-meter deep stone cave is situated behind the main hall. In the cave are unusual black and white bead-shaped stones, which this temple was named after. On the cave stands the Amitabha Hall where an Overlooking Pavilion stands in the front.

▲ The Goddess of Mercy Hall

The eighth Great Site of the Badachu Park is the Zhengguo Temple, which is located at the Lushi Hill and opposite to the other seven temples over the hill. The temple was first constructed under the reign of the Sui Emperor Renshou and was originally named the Shituo Temple. It is the oldest temple among all the eight Great Sites.

The south-facing Zhengguo Temple is divided into three building groups. The middle group from south to north is the Main Gate Hall, the Hall of the Heavenly Kings and the Hall of the Buddha of the Past, the Present and the Future. The Abbot Hall lies in the east and the Mimo Rock (the Secret Demon Rock) and other annexes stand in the west. Mimo Rock is a huge rock stretching out over the peak of the hill and suspended in the midair. Bending downwards, the rock was engraved with fours characters "Tian Ran You Shi" (A Picturesque Natural Rock). Under the huge rock lies a room made of stone where Monk Lushi lived and practiced Buddhism in the Sui and Tang Dynasties according to the legendary story. This hill was, therefore, named the Lushi Hill.

▲ The Secret Demon Rock

▲ The Main Gate Hall

▲ The Hall of Buddha of the Past, the Present and the Future

Buddhist Temples

063

# The Shangfang Mountain Temples and the Yunshui Cave

Located in Shangfang Mountain National Forest Park in Hancunhe County in Fangshan District, the Shangfang Mountain Temples and the Yunshui Cave has been a Buddhist place of interest since the Sui and Tang Dynasties. It was designated by Beijing Municipal Government as major historical and cultural site to be protected in 1984.

Known as Liupin Mountain in the past, Shangfang Mountain is the branch range of the Dafang Mountains. With the peak as high as 880 meters, the Shangfang Mountain is precarious. With luxuriant cypress trees, the environment is peaceful and beautiful. There used to be 9 caves, 12 peaks and 72 temples and nunneries represented by the Doushuai Temple. However, only 16 temples and nunneries remain till now for lack of renovation.

The Jiedai Nunnery (Reception Nunnery) at the foot of the Shangfang

▲ The Ladder in Cloud

▲ The Ladder in Cloud Nunnery

▲ Overlook of the Doushuai Temple

▲ The Hall of the Heavenly Kings

Mountain is the starting place for visitors who are going to climb the mountain. Walk through the Jiedai Nunnery, and proceed to the north, travelers will see an entrance to the mountain. Walk through the entrance and then make a turn, and then walk past the Fahan Mountain, travelers stay on the road until they reach the passage to the Yunshui Cave—Ladder in Cloud. On both sides of the Ladder in Cloud are fastened iron chains for visitors to hold. The stair of steps is high and steep. The Ladder in Cloud was built against the mountain on one side and

Buddhist Temples

065

against a short wall for protection on the other side. At the far end of the Ladder in Cloud is situated the Yunti Nunnery (Ladder in Cloud Nunnery). Small as it is, the nunnery is a must passage to the mountain. Several Taoist statues in the Ming Dynasty are preserved in the nunnery. Standing in front of the nunnery, travelers can have a bird's-eye view of the mountain.  Walk out of the Yunti Nunnery and walk north along the road and past the Kuanlong Bridge, the travelers will reach Doushuai Temple, the largest temple among all of the 72 nunneries and temples in Shangfang Mountain. Built between the Sui and Tang Dynasties, the temple underwent renovations in the years under Emperor Jiajing and Wangli's reigns in the Ming Dynasties. In front of the temple were built three stone steles in the Ming Dynasty, on which the history of the temple was recorded. On the rear wall behind the main hall were embedded 15 boards on which 42 buddhist scriptures were inscribed. In a courtyard which was located southwest to the temple, a number of tomb pagodas were built to bury monks' remains. Among them, a stone pagoda is the oldest one, built over 900 years ago in the Liao Dynasty. Opposite to the

▲ The Mahavira Hall

▲ Buddha Statues in the Mahavira Hall

temple, a little slight to the west, is the Zhaixingtuo Peak (Star-Picking Peak) which is the highest peak in the Shangfang Mountain. Beside the temple, other well preserved structures include the Scriptures-Storing Pavilion which was built in the Republic China period and the Wanghai Nunnery (Sea Watching Nunnery), etc.

Walk out of the Doushuai Temple and travel southwest for four kilometers, the travelers will reach the Yunshui Cave which is natural limestone cave, a wonder in Shangfang Mountain.

▲ Buddha Statues in the Stupa Pagoda

▲ Pagoda Forest in Pagoda Yard

Buddhist Temples

067

# The Cishan Temple

The Cishan Temple is located on the Tiantai Mountain at Wulituo in Shijingshan District, Beijing. There is a terrace in the west of the main peak, so the terrace is named Tiantai Terrace. Located on the terrace, Cishan Temple is named Tiantai Temple. It was designated by Beijing Municipal Government as a major historical and cultural site to be protected in 1995.

There are no definite records about when the Cishan Temple was constructed. There are two different opinions. One supports that the temple was built in the Qing Dynasty and the other shows that the temple was built in the Ming Dynasty. The iron bell in the temple was cast with the characters "Tian Tai Shan Ci Shan Si, Kang Xi San Shi San Nian" (it means "the Tiantai Mountain Cishan Temple, in the 33rd year in Emperor Kangxi's reign"). All the inscriptions on the steles existent were all inscribed in the Qing Dynasty, which indicates that the temple was completed at the beginning of the Qing Dynasty at the latest. Beijing Municipal Government invested to conduct a complete renovation from 2004 to 2006. Now, the front and rear hall in the present Cishan Temple are all newly renovated and most of the statues of Buddha have been remolded. It is open to the public as a scenic spot in Shijingshan District.

The south-facing Cishan Temple is composed of three building groups from the front to the back. It consists of the halls in the front of the temple gate, the hillside of the east mountain and the temple, which forms a temple layout that combines Buddhism, Taoism and the folk immortals at the same time. Constructed on the mountain, the Cishan Temple has 38 temples and over 150 rooms.

Outside of the gate of the first building groups stand the Wenchang Pavilion, the Jieyin Hall and the Hall of the Reclining Buddha.

On the east side of the mountain roadside from south to north are laid out several temple halls: the Goddess of Mercy Pavilion, the Small God of Wealth Temple, the Grandma Wang Hall, the Maitreya Hall, the Holy Spring, the Dragon King Temple, the Niangniang Temple, the God of Fire Hall, the Lvzu Hall (the Lv Dongbin Hall), the God of Horse Hall. On the east side of the mountain erect the Lantern Pagoda, the Arhat Cliff and the God of Mountains Temple. The Temple fairs held here in the past had the east, west, south and north God Greeting Procession. Prayers from various places gathered here. The temple fair was more than 300 meters long and decorated with a lantern every ten steps, which was fairly spectacular. On the south mountain slope erects a ten-meter high pagoda in an inverted-alms-bowl style.

The main gate of the south-facing Cishan Temple is the Skanda Hall. The Buddhist temples in the west of the Cishan Temple comprise two sets of courtyards.

▲ The Wenchang Pavilion

▲ The Goddess of Mercy Pavilion

The main hall in the forecourt is the three-bay Great Mercy Altar with a corridor in the front and a veranda in the back. The gold-lacquered and wood-carved statue of the thousand-hand Bodhisattva is enshrined in the middle of the temple. On both sides of the statue are offered eight statues of Gods such as the statue of God Bixia. Temples that contain both Buddhism and Taoism are rarely seen in Beijing. The east side hall is the Garan Hall and the west side hall is the Ksitigarbha Hall where the Ksitigarbha Buddha is enshrined. The main hall in the backyard is the two-story Pavilion for Storing Buddhist Scriptures. The statue of Monk Prince of the Devils is offered on the lower level and the Buddha of the Past, the Present and the Future is enshrined on the upper level. The east side hall is the Yuantong Hall and the west side hall is the Bodhidharma Hall where Bodhidharma and Goddess of Mercy are worshiped. On the west side of the Hall for Storing Buddhist Scriptures were built three north rooms where the famous patriotic General Feng Yuxiang lived when he went sightseeing and enjoyed recuperation on the Tiantai Mountain three times from 1912 to 1925. During his stay in the temple, he wrote down the following words in regular scripts on six spots: "Qin Jian Wei Bao" (Hardworking and Thrifty Is Treasure), "Zhen Chi Ku" (Bear Hardships) , "Geng Du" (Farming and Studying), "Dan Bo" (Not to Seek Fame and Wealth), "Ling Jing" (Area with Spectacular Landscapes) and "Qian Gua" (Modesty). His scripts were engraved on

▲ The Small God of Wealth Temple

the east mountain slope in the front of the temple main gate and on the north mountain slope behind the temple and they are all well preserved now.

In the east of the Cishan Temple are laid out five Taoist temple courtyards. The main hall in the forecourt is the San Huang Hall (Hall of Three Emperors). The east side hall is the God of Wealth Hall (The Monk Spring lies in the north) behind which comes the Abstinence Hall. The Lvzu Hall comes afterwards and is flanked with two side halls on each of its two sides. The last structure is the Hall for Storing Buddhist Scriptures. On the mountaintop in the front of the Cishan Temple are laid out the Hall of the Jade Emperor, the Tianqi Temple and the God of Mountains Temple.

▲ Tibetan-style Pagoda Covered with a Bowl-like Top

▲ The Great Mercy Altar

▲ Dragon-Patterns Ridge Ornament of Great Mercy Altar

▲ Phoenix-Patterns Ridge Ornament of Great Mercy Altar

# The Biyun Temple

Located at the eastern foot of the Beijing Fragrant Mountain,Haidian District, the Biyun Temple (The Azure Clouds of Temple) is one of the famous temples on the Western Hills in Beijing. The environment in the temple is quiet and beautiful and the structures are laid out geometrically. The Throne Pagoda and the statues of the 500 Arhats are unsurpassed. It was designated by the State Council as the major historical and cultural site under state protection in 2001.

The Biyun Temple was first constructed as a mansion in the 2nd year (1331) in the reign of the Yuan Emperor Zhishun. Yelv Chucai's descendant Yelv Aji transformed it to a temple and named it the Biyun Nunnery and then renamed it the Biyun Temple. The temple was renovated in both the Ming and Qing Dynasties. During the period of Ming Emperor Zhengde, Eunuch Yu Jing was satisfied with this geomantic treasure land and hence began a large-scaled construction which was regarded as the first extension. He also prepared a tomb for himself behind the temple. Yu Jing committed a crime in the 1st year in the reign of the Ming Emperor Jiajing and could not be buried here. Eunuch Wei Zhongxian expanded

▲ Old Picture of the Biyun Temple (Taken in 1920s)

Buddhist Temples

▲ Panoramic View of the Biyun Temple

the temple in the reign of the Ming Emperor Tianqi and set up a tomb for himself once again with the intention of being buried here after he died. After Wei Zhongxian took a suicide, his dead body was exposed to the public and consequently could not be buried here. Ge Jiusi, the confederate of Wei Zhongxian entered Beijing with the Qing Army and buried Wei's garment in the tomb hence it became Wei's cenotaph.Until the 40th year in the reign of the Emperor Kangxi, the imperial supervisor of inspecting roads found the tomb when he received the order to inspect the Western Hills. He considered it as the imperial mausoleum of the previous dynasty at the very beginning and then got to know the truth. Consequently, he reported it to the emperor on May 12 and asked for the permission to raze the tomb to the ground on May 22. The temple was renovated in the 13th year (1748) in the reign of the Emperor Qianlong and a Throne Pagoda was built according to the architectural drawing provided by the monks in the temple, besides, new temporary imperial palaces and an Arhat hall were also built. No large-scaled renovations were conducted to the other structures so the halls in the temples were basically constructed in the Ming Dynasty.

Sun Yat-sen's coffin had stayed in the rear hall of the temple after he died in Beijing in 1925, so the hall was renamed (the Sun Yat-sen Memorial Hall). The Throne Pagoda had become the cenotaph of Sun Yat-sen. The monument to Sun Yat-sen beside the street was transferred to the open space before the Throne Pagoda in the temple because the street before the east gate of the Summer Palace had to be broadened in 1983.

The east-facing Biyun temple houses three complexes, which include the main structures in the middle, the spring courtyard structures and the Hall of the Arhats. The main halls are among the structures in the middle. In the front of the main halls stand a stone bridge and a pair of Ming stone lions carved with white marbles standing on the east end of it. The main gate to the temple is located behind the bridge and is the first entrance of the hall. It is beamless and single-eaved with a flush gable roof covered with round gray tiles and supported by a main ridge. Both sides of the gate were connected by a two-story watchtower-styled structure with a gable-and-hip roof covered with gray round tiles and tiger-skin-colored stone metope. The Main Gate Hall has three rooms and is the main hall of the first courtyard. It has a single-eaved hip roof covered with round gray tiles and decorated with colored tangent circle patterns. The Un and Ah Generals are offered in the hall. The doors were installed in the wall on both the left and right sides of the hall. The Maitreya Hall is the main hall in the second courtyard that is three bays wide with a single-eaved gable-and-hip roof covered with round gray tiles and decorated with colored tangent circle patterns. The Maitreya Buddha is offered in the hall. In the courtyard stands a Bell Tower and a Drum Tower. Located behind the Maitreya Hall,

▲ Stone Bridge

▲ The Main Gate Hall

▲ The Mahavira Hall

the Mahavira Hall is three bays wide and with a hip roof in a single-eaved style and covered with round gray tiles and carrying a gallery. Under the eave were decorated with colored tangent circle patterns. There is an outer room at the back eave with an annex decorated with colored tangent circle patterns. Sakyamuni Buddha is offered in the hall. There is a Dhanari Column on both the south and north sides of the platform before the hall. Meanwhile three side halls stand on each of the south and north sides of the hall. Behind the Mahavira Hall erects the Bodhisattva Hall which is five bays wide and surrounded by galleries. Bodhisattvas are offered in the hall. On the platform before the hall stands a two-story hexagonal pavilion built over a stele with a pavilion roof. Behind the Bodhisattva Hall was built the Sun Yat-sen Memorial Hall, which is five bays wide with a single-eaved gable-and-hip roof covered with round gray tiles. Ceilings under the eave were decorated with colored tangent circle patterns. The coffin of Sun Yat-sen once had a short stay here. Three auxiliary halls are annexed to the Sun Yat-sen Memorial Hall that is used for exhibiting Mr. Sun's life story now. The Throne Pagoda is situated far in the back of the middle structure group. A wooden memorial archway was set up in the front of the pagoda with a stone archway and brick masonry archway as the guide, which sets off the mystery and solemnity of the pagoda. Built by magnificent white marbles, the 34.7 centimeter tall pagoda was founded on a two-tiered tiger's skin-patterned stone base and was divided into the upper and lower levels. The Vajrasana Pagoda is seated in the lower level and a door with caves situated in the middle, where the garment of Sun Yat-sen was buried. Walking to the pedestal along the stone steps, visitors can find a square pavilion near the exit where stands a Tibetan-

▲ Sunk Panel in the Mahavira Hall

▲ The Stone Pailou

styled pagoda with an inverted-alms-bowl roof on both the left and right sides. Behind the pavilion stand five thirteen-storied square pagodas in a multi-eaved style among which the main pagoda is located in the center and the other four around it are the side pagodas. The whole Vajrasana Pagoda was covered with delicate relieves including the big and small Buddha statues, the Heavenly Kings, the Strong Man, the dragon, phoenix, lion and elephant designs and the cloud-shaped flowers, which are the exquisite stone carvings during the period of Qing Emperor Qianlong.

In the northern side courtyard of the Biyun Temple is found the Spring Garden which is an imperial palace for a short stay combining the characteristics of garden and imperial residence. The whole courtyard is east-facing and the first three structures house the Azure Hall of Study complex and the last two are the Spring Gardens.

The Hall of the Arhats (Luohantang), imitating the Jingci Temple in Hangzhou, was built in the 13th year during the period of Qing Emperor Qianlong. The east-facing hall is located at the South Biyun Temple Road and is nine bays wide with a hip roof covered with round gray tiles. An inverted-alms-bowl-styled pagoda is

▲ The Sun Yat-sen Memorial Hall

▲ The Vajrasana Pagoda

▲ The Azure Hall of Study

▲ Panoramic View of the Spring Garden

Buddhist Temples

079

▲ Arhats in the Hall of Arhats (No.1)

located on the main ridge. There are three verandas stretching out of the front eave with a gable-and-hip roof. Under the eave of the outer room were inscribed with the golden characters "Luo Han Tang" (The Hall of the Arhats) on a crouching board. The Four Heavenly Kings are enshrined in the main gate. There are 508 Arhats inside with various expressions, showing superb carving skills. They are considered to be unique of the Biyun Temple. It was said that Emperor Kangxi and Qianlong were among the Arhats. The Buddhas of the Past, the Present and the Future are seated in the center of the hall and a Buddha stands in each of the passages in the four directions. Vajra Dharma Skanda stands in the east, Buddha Mad Monk is in the north, Bodhisattva Ksitigarbha lies in the west and the Ambassador Buddha stands in the south. The living Buddha Jigong is seated on the north beam of the hall.

Buddhist Temples

▲ The Hall of Arhats

081

▲ Panoramic View of the Hall of Arhats

▲ Arhats in the Hall of Arhats (No.2)

# The Wanshou Temple

The Wanshou Temple (The Temple of Longevity) is located in the northeast of Zizhu Bridge on the West Third Ring Road, Haidian District in the downtown areas of Beijing and is one of the famous temples in western Beijing. The temple was an imperial palace for the imperial family to have short stays during the Ming and Qing Dynasties. It was designated by the State Council as the major historical and cultural site under state protection in 2006.

Originally named the Juse Temple, the Wanshou Temple was first built in the Ming Dynasty. It was renovated in the 5th year (1577) under the reign of Emperor Wanli in the Ming Dynasty and was renamed Wanshou Temple which was mainly used for storing Buddhist scriptures. After the word-carved boards and the Buddhist scriptures were transferred to the Fan Blocking-printing Atelier and Han Blocking-printing Atelier (the Fayuan Temple), the Wanshou Temple later became a place for the imperial families of the Ming and Qing Dynasties to have a rest on the tour to the West Lake (the Kunming Lake). It was destroyed by the war fire in the Ming Dynasty and got renovated in the 25th year (1686) under the reign of the Qing Emperor Kangxi. Emperor Qianlong celebrated his mother's birthday in the temple in the 16th year (1751) and the 26th year (1761). So it underwent two renovations and the imperial halls in the west of the temple were added to the temple. It was

▲ The Main Gate

▲ The Mahavira Hall and the East Side Hall

destroyed by fire in the 1st year in Emperor Guangxu's reign and later became a vegetable garden. In order to celebrate the birthday of Empress Dowager Cixi, the temple was renovated once again in the 20th year (1894) under the reign of Emperor Guangxu. The Qianfo Pavilion and the Dressing Tower were added to the west courtyard. Plus the vegetable garden, the final framework was formed. Emperor Qianlong celebrated his mother's birthday three times in the temple and Empress Dowager Cixi worshipped the Buddha in the Wanshou Temple and had some refreshments in the west side yards hence the temple was also called the small Palace of Peace and Longevity. In the year about 1934, the front courtyard of the temple once served as a school for the refugees' children from northeast of China. The Beijing Art Museum was set up in the middle courtyard in 1985. It is famous for collecting Ming and Qing artworks.

With an area of 31800 square meters, the Wanshou Temple complex is divided into three courtyards. The middle courtyard is the main part of the temple, the west courtyard contains the imperial halls and the east yard is the abbot halls. The middle courtyard has eight halls that laid out orderly from the south to north are the main gate with three rooms and a gable-and-hip roof covered with round tiles. Bestowed in the 2nd year in Emperor Shunzhi's reign in the Qing Dynasty, the stone board hung above its door inscribed with the characters "Ci Jian Hu Guo Wan Shou Si" (Construction of the Huguo Wanshou Temple under imperial command). The Hall of the Heavenly Kings has three rooms and a gable-and-hip roof covered with round tiles. In the hall are enshrined the Maitreya Buddha, the Guardian Deities and the images of the Four Heavenly Kings (non-existent now). The Bell Tower and Drum Towers are erected respectively on the east and west side of the Heavenly Kings

Buddhist Temples

▲ The Wanshou Pavilion

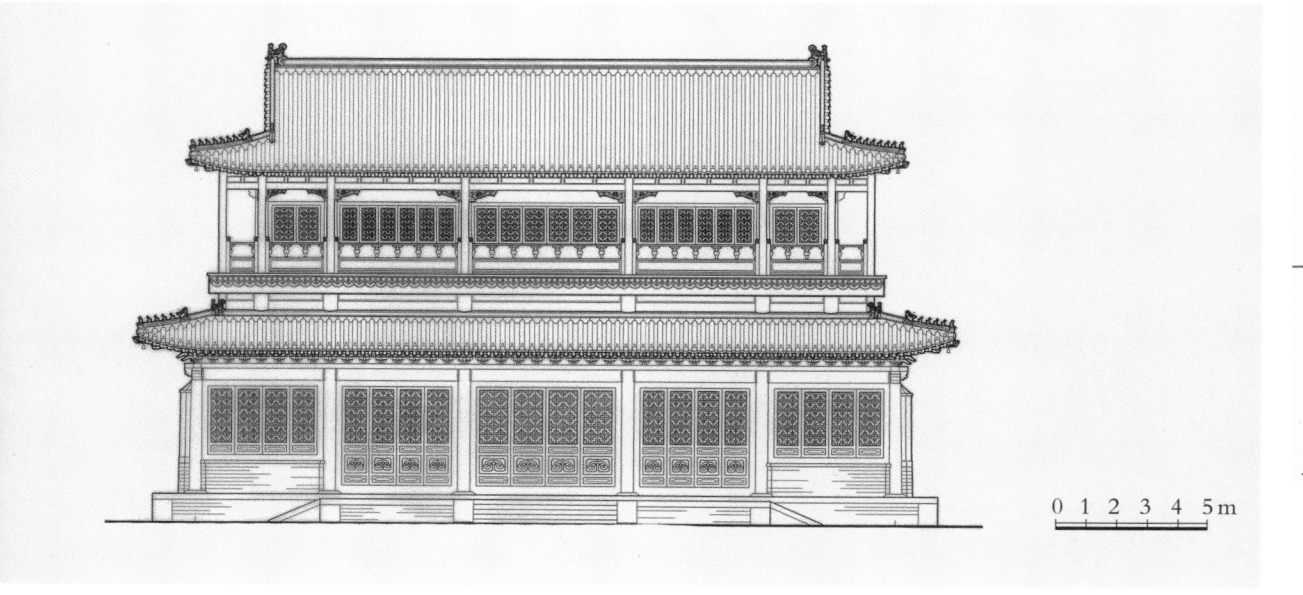

0 1 2 3 4 5m

▲ South Elevation of the Wanshou Pavilion

085

Hall. They have a double-eaved gable-and-hip roof covered with round tiles and the "Big Yongle Bell" was once positioned here (moved to the Big Bell Temple later). The Mahavira Hall is five bays wide with a hip roof covered with glazed tiles. The Pilu Buddha and the Buddha of the Past , the Present and the Future are enshrined here. The Mahavira Hall has three side halls on both its east and west sides with gable-and-hip roofs covered with round tiles. The east side hall is called "Longevity Wish" and the west side hall is called "Inner Peace Mirror".  The Wanshou Pavilion was also named the Ning'an Pavilion (the Peace Pavilion). It is seven bays wide with a gable-and-hip roof covered with round tiles and both the east and west side halls have three rooms. In the east comes the Skanda Hall and in the west stands the Bodhidharma Hall. Both of them have a gable-and-hip roof covered with round tiles. There is a big meditation hall with five rooms used by the monks to sit in meditation. It has a flush gable roof covered with round tiles and suspended by a main ridge. The east and west halls are small meditation halls all used for the monks to sit in meditation. The fifth courtyard is very unique and was constructed on an artificial hill. The Goddess of Mercy Hall stands in the center; the Manjusri Hall is located in the east and the Samantabhadra Hall erects in the west, which indicates Putuo Mount, Qingliang Mount (Wutai Mount) and Emei Mount, the three Buddhist

▲ Ornamental Windows on Both Sides of the Western-style Gate

▲ The Goddess of Mercy Hall

▲ The Qianlong Imperial Stele Pavilion in the Qing Dynasty

▲ The Western-style Gate

▲ The Wanfo Tower

Holy Mountains in China. The pavilion built over a stele under Emperor Qianlong's command has double eaves and an octangular pavilion roof covered with glazed yellow tiles. In the center erects a stele inscribed with the words "Chong Xiu Wan Shou Si" (Renovate the Wanshou Temple) by Emperor Qianlong. The inscription was written in four languages: Chinese, Manchu, Mongolian and Tibetan. The one-bay Amitayus Buddha Hall has a double-eaved gable-and-hip roof covered with round tiles. Two Baroque-style doors were installed on the two sides of the Amitayus Buddha Hall and they were built in the 26th year (1761) in Emperor Qianlong's reign with a very unique style among all the imperial temples. The Imperial Stele Pavilion built under Emperor Guangxu's command has a double-eaved octangular pavilion roof covered with glazed yellow tiles. The pavilion houses Weng Tonghe's inscription on the stele set up in the 20th year (1894) in the reign of Emperor Guangxu. The last structure in the middle courtyard is the two-story Wanfo Building that is seven bays wide with a flush gable roof covered with round tiles and supported by a main ridge. There are three side halls in front of the building.

▲ Transverse Section of the Western Road of the Wanshou Temple

The east courtyard houses the Abbot Hall serving as venues for the abbots to study Buddhist scriptures and as their residences.

The west courtyard contains the imperial halls among which are the Gate, the Birthday Celebration and Refreshment Room, the Front Main Hall, the Main Hall, the Dressing Tower and the Hall of Great Mercy etc. The structures own both an imperial style and characteristics of a garden. Empress Dowager Cixi used to take a rest and drink tea here when touring around the Summer Palace.

▲ Tea House

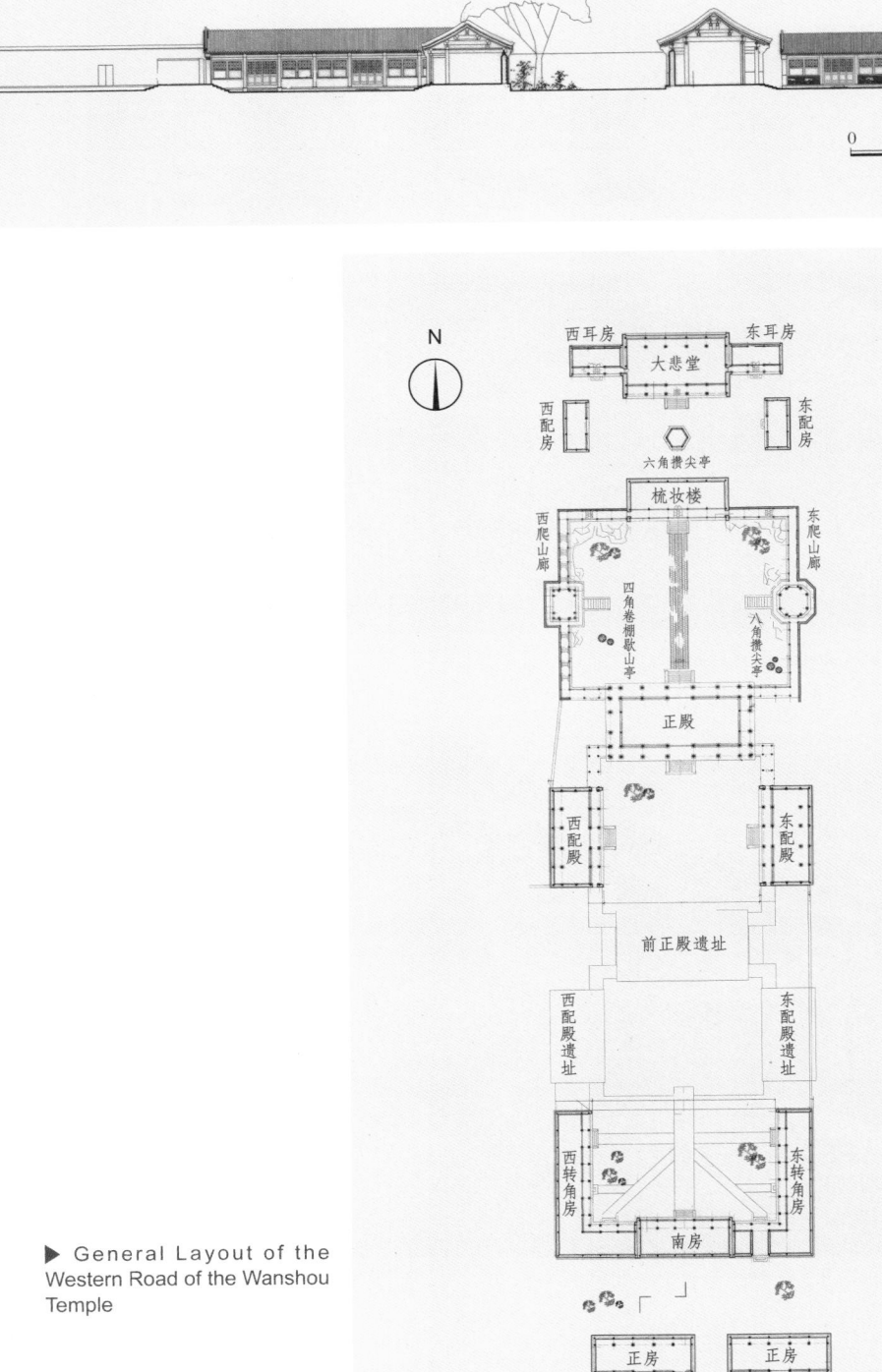

Buddhist Temples

▶ General Layout of the
Western Road of the Wanshou
Temple

0  5  10  15 m

0 5 10 15 20 25 m

▲ Dressing Tower

▲ The Open Hall and the Gallery

# The Baoguo Temple

The Baoguo Temple (The Temple of Longevity) is located at No. 1 Baoguosi Front Street in Xuanwu District, Beijing. It was designated by the State Council as a major historical and cultural site under state protection in 2006.

The Baoguo Temple was first constructed in the Liao Dynasty and was renovated in the 2nd year of Chenghua's reign (1466) in the Ming Dynasty. Zhou Jixiang, the younger brother of Queen Mother Zhou, became a Buddhist and practiced Buddhism in this temple. Most of the architectures in the Baoguo Temple fell down during the earthquake in Beijing in the 18th year (1679) during Kangxi's reign. It was renovated and renamed "Da Bao Guo Ci Ren Si" in the 19th year during Emperor Qianlong's reign in the Qing Dynasty. After the *Xin Chou Treaty* was singed, Zhaozhong Ancestral Hall was included in the Embassy District in the Dongjiaominxiang. The Qing Dynasty had planned to transform the Baoguo Temple into the new Zhaozhong Ancestral Hall, but this plan did not work out due to the extinction of the Qing Dynasty.

The south-facing Baoguo Temple has four courtyards. The main gate is three bays wide and five purlins deep with a corridor in the front of it. It has a flush gable roof supported by a main ridge and covered with round tiles. The main hall of the

▲ The Main Hall of the Second Courtyard

first courtyard is five bays wide and seven purlins deep with an overhanging gable roof. A stele erects on each of the eastern and western side before the hall. Behind the main gate, visitors can see the second courtyard whose main hall is three bays wide and seven purlins deep with an overhanging gable roof. A stele erects on each of the eastern and western side before the hall. The second courtyard has five side halls on each of its east and west side. The main hall in the third courtyard is nine bays wide and seven purlins deep with a gable-and-hip roof covered with round gray tiles and carrying sheared green edges. In the front of the hall was built a wide brick platform. The main hall has five side halls on each of its east and west side. The main hall in the fourth courtyard is five bays wide and seven purlins deep. It is flanked with three side halls on each of its two sides and has three side halls on each of its east and west side.

▲ Gable Wall of the Second Courtyard

▲ The Main Hall of the Third Courtyard

▲ Gallery in the Gu Tinglin Ancestral Hall

▲ Stone Stele in the Gu Tinglin Ancestral Hall

The Gu Tinglin Ancestral Hall is located at the south end of the West Baoguosi Road. Gu Yanwu (also styled Tinglin), a famous scholar in the Qing Dynasty, once lived in the hall. The Gu Tinglin Hall was destroyed by the Eight-Power Allied Forces in 1900 and was renovated by Wang Shitong etc. in 1921. The Gu Ancestral Hall was once occupied by Zhixing Middle School in 1934.

Gu Tinglin Ancestral Hall served as a warehouse afterwards and its original site was severely destroyed. After the liberation of China, the Gu Ancestral Hall and the Baoguo Temple were occupied successively. The Beijing Municipal Government contributed to renovating the Baoguo Temple and the Gu Tinglin Ancestral Hall in 1990.

The south-facing Gu Tinglin Ancestral Hall has three courtyards. On the southeast corner of the courtyards was built a brick bar. The stone board on the arch was inscribed with the seal scripts "Gu Xian Sheng Ci" (the Ancestral Hall for Mr. Gu). Each courtyard is composed of the main halls and the west side halls which all possess their own unique style. The courtyards are connected by a veranda from south to north. The first courtyard has five main rooms and three south rooms with a west side room. With the same layout, the second and third courtyards have three main rooms flanked with a west wing room and three west side rooms inside the courtyards. All the architectures in the courtyard have a flush gable roof supported by a raised ridge and covered with valvate tiles.

# The Guangji Temple

Situated at No. 25 Fuchengmennei Street in Xicheng District of Beijing, the Guangji Temple is now the head office of the Chinese Buddhism Association. It was designated by the State Council as a major historical and cultural site under state protection in 2006.

The Jin Dynasty saw the construction of the Guangji Temple which was originally named the Xiliucun Temple. It was renamed "the Return Hongji Temple" in the Yuan Dynasty and was destroyed by war at the end of the Yuan Dynasty. In the years of the Tianshun reign in the Ming Dynasty, Puhui and Yuanhong, two monks from Shanxi Province, came here to raise funds and rebuild this temple at the same site. It was finally completed after two years. Emperor Zhu Jianshen ordered to name it "the Hongci Guangji Temple" in the 2nd year (1466) of his reign. It experienced expansions in the 12th year (1584) of Emperor Wanli's reign in the Ming Dynasty and in the 33rd year (1694) of Kangxi's reign in the Qing Dynasty. It was renovated in the 38th year (1699) and an imperial stele and the gold-glided statue of Sakyamuni were added to the temple. It was destroyed by fire in 1921 and rebuilt in 1924. After suffering a severe fire disaster in January 1934, the Main Hall and the Back Hall in the temple were burned up. The classical scriptures and the

▲ The Main Gate

▲ The Mahavira Hall

sandalwood-made statue of Sakyamuni were all burned. The temple was renovated
in the next year (1935). It went through three large-scale renovations in 1952, 1972
and 2000.

The south-facing Guangji Temple covers an area of 2.33 hectares. On its
central axis stand four important buildings: the Main Gate, the Hall of the Heavenly
Kings, the Mahavira Hall, the Yuantong Hall (the Buddhisattva Hall) and the Hall
for Storing Buddhist Scriptures (the Stupa Hall) etc. Besides, in the west courtyard
stands the three-story white marble Ordination Altar.

The Guangji Temple contains three main gates joined to each other by
connecting walls. The middle main gate has a gable-and-hip roof covered with
glazed green tiles and carrying with sheared yellow edges and it also has a stone
arched doorway above which hangs a lintel inscribed with the characters "Chi
Ci Hong Ci Guang Ji Si" (Name of the Hongci Guangji Temple Given by the
Emperor). The east and west doorways have the same design as the middle gate
but in a much smaller size. The east lintel was inscribed with the characters "Di Lu
Xing Hai" and the west lintel was inscribed with the characters "Hua Zang Xuan
Men". On both sides of the gate stands a splay screen hall. Inwards, visitors will see
the Bell Tower and the Drum Tower on the left and right side of the main gate. The
Hall of the Heavenly Kings in the middle is three bays wide with a gable-and-hip
roof covered with round tiles. An arched stone doorway was installed in the temple
in which the bronze seated statue of the Maitreya Buddha is enshrined and guarded
by the Four Heavenly Kings on the left and right sides. The statue of Skanda is
offered at the back of the Maitreya Buddha. In the second courtyard is the five-
bay Mahavira Hall with a hip roof covered with glazed round yellow tiles. In the

▲ Part of the Finger Drawing *The Shengguo Ghosa Painting* of the Qing Dynasty in the Mahavira Hall

front of the hall is a bronze vessel that was molded in the 58th year (1793) under the reign of Emperor Qianlong in the Qing Dynasty. This fine vessel is about two meters high cast with the eight treasures of Buddhism (the wheel, the spiral shell, the umbrella, the canopy, the flower, the vase, the fish, and the knot). The designs on the vessel are simple and sophisticated and carved with exquisite craft, making the vessel a precious art treasure. Inside the hall are enshrined the statue of the Buddha of the Past, the Present and the Future and the bronze statues of the Eighteen Arhats. Across their backs is a finger painting named *The Shengguo Ghosa Painting* painted by the Qing artist Pu Wen. The painting is five meters high and ten meters wide, depicting the scene when Sakyamuni Buddha is expounding Buddhist doctrines. The Yuantong Hall in the third courtyard is the residence of Avalokitesvara Bodhisattva. It is five bays wide with a flush gable roof supported by a main ridge and covered with round tiles. The fourth courtyard houses the two-story Rearmost Hall and on the upper part stands the Stupa Hall with a glazed green-tiled roof. On the lower part stands the Treasure Hall with a glazed yellow-tiled roof. It displays the precious curiosities that have been given as gifts by the friendly Buddhists from other states. The tooth relic of the Buddha of the Lingguang Temple was enshrined in the Stupa Hall from 1955 to 1964. The hall is now a place where Buddhist scriptures are stored. More than one hundred thousand volumes of Buddhist scriptures are stored and the rubbings of the stone-carved scriptures from the Yunju Temple are also preserved here. It also stores a volume of *Tripitaka* printed by the printing blocks in the Zhuoni Temple at Lintan County in Gansu province, totaling 231 packages which are a rare version among all the Buddhist collections. Among them, what is particularly precious is the Buddhist scripture written with blood in the Song and Ming Dynasties. The west courtyard houses the Chilv Hall, the Jingye Hall and the Yunshui Hall (the Cloud and Water Hall).

In the northwest corner of the temple are situated the well-preserved Mandala Hall and the white marble Mandala Altar that is called the "Sanxue Hall" today. There is an ancient tree in the temple and the stele beside the tree was inscribed with the prose entitled *The Song to the Sago Cycas* written by Emperor Qianlong himself.

▲ The Yuantong Hall

▲ The Pavilion for Storing Buddhist Scriptures

▲ The Sanxue Hall

# The Bailin Temple

Located at No.1 Xilou Lane in Dongcheng District of Beijing, the Bailin Temple is distinguished for preserving the Dragon Tripitaka. It was designated by the State Council as the major historical and cultural site under state protection in 2006.

The Bailin Temple was first constructed in the Yuan Dynasty and completed in the 7th Zhizheng year of the Yuan Dynasty (1347). It was renovated in the 12th year (1447) in Emperor Zhengtong's reign in the Ming Dynasty and the Li Shimian, the Educational Administration Executive in the Imperial Academy, inscribed the epigraphy for renovating the temple. The largest scaled renovation was conducted in the 52nd year (1713) in Emperor Kangxi's reign. In order to celebrate the 60th birthday of Emperor Kangxi, Prince Yinzhen, the successive Emperor Yongzheng, was in charge of the large-scaled unprecedented renovation.

The five main structures in the temple compound are laid out on a central axis from south to north. Listed from the south to the north of the temple, they are as follows: the main gate, the Devaraja Hall (the Hall of Heavenly Kings), the Mahavira Hall, the Amitabha Hall and the Hall of the Ten-thousand Buddhas. These structures are all flanked with side halls and galleries in the east and west. The temporary imperial palaces, the Abstinence Hall and the small Dharma Hall are situated in the east and west of the temple. With an orderly and rigorous layout,

▲ The Screen Wall

▲ Brick Relief Sculpture on the Screen Wall

▲ The Hall of the Heavenly Kings

Buddhist Temples

all the structures of the temple were built on a tall brick base, making the whole temple imposing and spectacular. In the front of the main gate erects a tall brick screen wall engraved with exquisite designs. As the main structure in the temple, the Mahavira Hall has double eaves and a gable-and-hip roof covered with round tiles. A horizontally inscribed board with the handwriting of Emperor Kangxi on his 60th birthday was hanged on the center under the eave of the Mahavira Hall. The inscription says "Wan Gu Bai Lin" (The Everlasting Cypress Grove). In the front of the hall was constructed a platform with a wall stretched out in the east, south and west side. Statues of the Buddhas of the Past, the Present and the Future are found inside the temple. Seven carved and gilded Buddha images dating from the Ming Dynasty are kept inside the hall. They are so vivid and life-like. To the east of the main hall is an auxiliary hall containing two large bronze bells of coiling dragons that are 2.6 meters tall cast in the 46th year (1707) in the reign of Emperor Kangxi. Their surfaces were delicately cast with exquisite scriptures. The Hall of the Ten-thousand Buddhas, also called the Hall of Vimalakirti, is located at the end of the temple and forms a single courtyard. The two-story hall was flanked with two buildings and planted with several tall and straight luxuriant cypresses, which creates a quiet and mysterious atmosphere for the courtyard.

Among the valuable relics in the temple is the only complete set of printing

▲ The Mahavira Hall

▲ Color Painting and Brackets in the Mahavira Hall

blocks for the Tripitaka existent today. The Buddhist scriptures were called "Tripitaka" ever since the Tang Dynasty. The Dragon Tripitaka refers to the collection of the Buddhist scriptures carved by imperial command in the Qing Dynasty. It includes an all-round content that collects the works by eminent senior monks and researchers of Buddhist scriptures in the Yuan, Ming and Qing Dynasties. Many historical data can be found in this collection, which is a treasure for studying culture, history, philosophy, art and translation. Costing six years,

▲ The Hall of Ten-thousand Buddhas

▲ The Abstinence Hall

Buddhist Temples

▲ Ornamental Gate in the Yard of the Imperial Temporary Dwelling Palace

the work of carving was begun in the 11th year (1733) during the reign of Emperor Yongzheng and was completed in the 3rd year (1738) during the reign of Emperor Qianlong. The collection has more than 7,000 volumes totaling more than 70,000 separate blocks. Carved of high-grade pear wood with delicate carving craft and clear and straight typeface and due to the rare printings, the characters in the blocks remain in fine condition today. The blocks were transferred to the Zhihua Temple in 1982.

▲ Rockwork in the Yard of the Imperial Temporary Dwelling Palace

# The Fahai Temple

Located at the south foot of the Cuiwei Mountain, the Fahai Temple is noted for its delicate frescos. It was designated by the State Council as the major historical and cultural site under state protection in 1988.

According to the inscription on the stele, Eunuch Li Tong and other eunuchs raised money for constructing the temple that was finally built by the Construction and Maintenance Section of the Ministry of Labor in the 4th year (1439) in the reign of Emperor Zhengtong. It was completed in the 8th year (1443) in the reign of Emperor Zhengtong and named by the Emperor Yingzong "the Fahai Buddhist Temple". According to the inscription on the stele, the Mahavira Hall was located in the center and the Garan Hall and the Hall of Patriarch stood on the left and right sides of the Mahavira Hall. Before the Mahavira Hall erected the Heavenly Kings Hall, the Bell Tower and Drum Tower. The Vajra Dharma Hall stood in the front of the Heavenly Kings Hall. The Pavilion for Storing Buddhist Scriptures erected behind the Mahavira Hall with an abbot room on its left and a Field for Electing the Buddhas on its right. Besides, there were hip-roofed halls, kitchens, etc. The

▲ Overlook of the Fahai Temple

temple was encircled by walls and a "far gate" was built in the south, about 500 meters away from the temple. After the completion of the temple, it went through a large-scale renovation from the 17th year (1504) in the reign of the Ming Emperor Hongzhi to the 1st year (1506) in the reign of Emperor Zhengde. It was renovated again in the 21st year (1682) in the reign of the Qing Emperor Kangxi. It underwent several renovations afterwards. The first renovation of the temple after the liberation was conducted in 1953. The main gate, the Mahavira Hall and the east and west hip-roofed halls were all renovated in 1982. The Hall of the Heavenly Kings was renovated in 1985 and the Mahavira Hall was rebuilt in 1995. The Pavilion for Storing Buddhist Scriptures was reconstructed in 2005.

The south-facing Fahai Temple is three bays wide with a gable-and-hip roof covered with glazed yellow tiles. Behind the main gate stands a rectangle compound stretching from the east to the west. There are tens of steps in the center of the compound to the Hall of the Heavenly Kings and the Mahavira Hall. On both sides of the steps stands a stele. On one stele was inscribed with the "Imperial Inscriptional Record for Constructing the Fahai Buddhist Temple" by Hu Ying, the Director of the Board of Rites in the 8th year (1443) under the reign of the Ming Emperor Zhengtong. On the other stele was inscribed with "Notes to the Fahai Temple" by Wang Zhi, the Director of the Ministry of Official Personnel Affairs. The three-bay Hall of the Heavenly Kings has a gable-and-hip roof covered with

▲ The Hall of the Heavenly Kings

▲ The Mahavira Hall

▲ Water and Moon Goddess of Great Mercy —— Wall Painting in the Mahavira Hall (No.1)

cut round tiles and carrying sheared yellow edges. The five-bay Mahavira Hall has a hip roof with sheared yellow edges and covered with cut round tiles. In the Mahavira Hall, there are delicate frescos. The frescos of the Goddess of Mercy (in the middle), Samantabhadra (on the left) and Manjusri (on the right) were painted on the back of the fan wall behind the Buddha and right facing the rear door. The two frescos on the two sides of the rear wall depicted Brahma worshiping the

▲ Wall Painting in the Mahavira Hall (No.2)

▲ Wall Painting in the Mahavira Hall (No.3)

Buddha. The frescos of Buddha and Bodhisattva were depicted on the east and west gable walls behind the statue of the Arhats. The frescos, especially the frescos of the Three Masters are exquisite keeping the artistic charms of the Ming Dynasty. This fresco is the most delicate and largest one extant in Beijing. Painted by skillful craftsman, the frescos in the temple top all the frescos extant in China and enjoy an important place in the Chinese history of frescos.

# The Zhihua Temple

The Zhihua Temple is located at No. 5 Lumicang Lane in Dongcheng District, Beijing. In the temple, the "Music of Beijing in the Zhihua Temple" is preserved, which is considered as "the Living Fossil of the Ancient Chinese Music". The temple was designated by the State Council as the major historical and cultural site under state protection in 1961.

The Zhihua Temple was first built on January 9 (lunar month) in the Chinese lunar calendar in the 9th year (1444) of Emperor Zhengtong's reign and was completed on March 1 (lunar month) in the same year according to the record of the "Notes to the Imperial Mandate to Construct Zhihua Temple" and the "Imperial Mandate to Set up a Bao'en Stele to Zhihua Temple". The Zhihua Temple

▲ The Main Gate

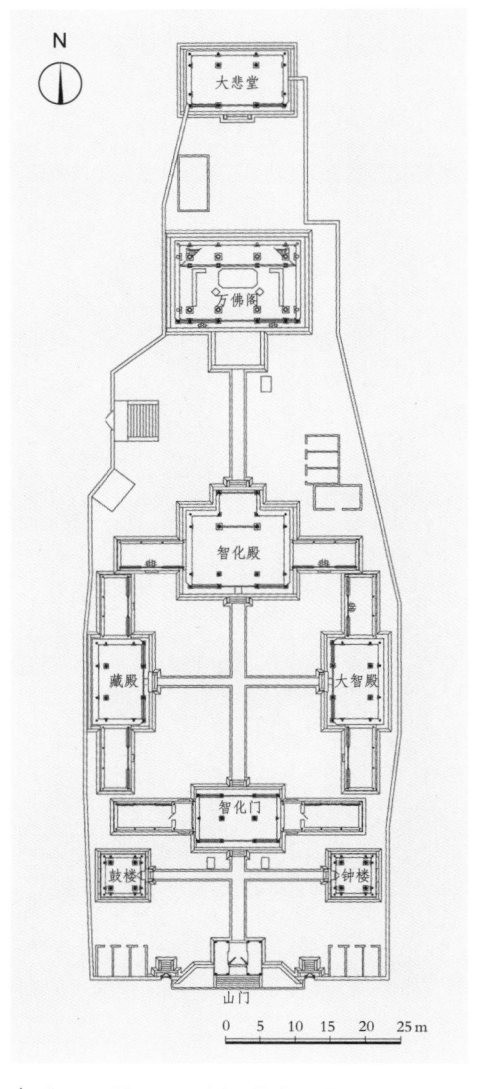

▲ General Layout of the Zhihua Temple

▲ Panoramic View of the Zhihua Temple

was formerly the ancestral temple of Wang Zhen, an eunuch of the Rites Supervising Office in the Ming Dynasty. And an ancestral hall named the Jingzhong Hall dedicated to Wang Zhen was constructed in the temple. In the 7th year (1742) in the reign of Emperor Qianlong, Messenger Shen Tingfang reported to Emperor Qianlong and asked for permission to destroy the Wang Zhen's Temple and the statue of Wang Zhen and pull down the stele for the purpose of "punishment". From then on the Zhihua Temple was on the decline and the heritages in the temple were gradually lost. The structures in the temple were not renovated at the end of the Qing Dynasty and in the Republic of China Period. After the liberation of China, in order to preserve the precious structures built in the Ming Dynasty, the government conducted several renovations and strengthened the management of the temple.

The south-facing Zhihua Temple has five courtyards and is one of the biggest complexes of temple buildings in the Ming Dynasty existent in Beijing. Liang Sicheng and Liu Dunzhen, experts in ancient architectures, both made specialized comments on the structures of the Zhihua Temple.

In a brick masonry structure in imitation of the timber structure,

Buddhist Temples

the three-bay main gate is one bay deep with a gable-and-hip roof covered with glazed round black tiles. The outer room had opened an arched doorway with a lintel inscribed with the characters "Chi Ci Zhi Hua Si" (Name it ZhiHua Temple by Imperial Command). Splay screen walls are built on the two sides of the main gate. Opposite to the main gate stood a brick screen wall pulled down after the liberation of China in 1949. From the temple gate inwards lays the first courtyard. On the east side of the temple stands the three-bay Zhihua Gate with a gable-and-hip roof covered with black glazed tiles. Situated respectively on the east and west side of the first courtyard, both the two-story Bell and Drum Towers had the same framework with a gable-and-hip roof covered with glazed black tiles. At the back of the Zhihua Gate stands the Zhihua Hall, the main hall of the temple, which is three bays wide with a gable-and-hip roof covered with glazed black tiles. Under the eaves was designed a bracket set whose architrave was decorated with colored tangent circle patterns. In the center of the main hall and main gate stands a white marble stone molded base. The statues of Sakyamuni, Amitabha and the Medicine Guru Buddha were enshrined in the center while the 18 seated Arhats were enshrined on the two sides. All the wooden statues were decorated with gold inlays. The Dazhi Hall (the Hall of Great Wisdom) is the east side hall of the Zhihua Hall and is three bays wide with a gable-and-hip roof covered with glazed round black tiles. The statues of Avalokitesvara, Manjuisri and Samantabhadra etc. were once enshrined here. The Tibetan Hall is the west side hall of the Zhihua Hall to which the Tibetan Hall has an identical framework. A revolving octagonal sutra cabinet

▲ The Zhihua Gate　　　　　　　　　▲ Buddha Statues in the Zhihua Hall

was once preserved in the Tibetan Hall. On the north side of the Zhihua Hall stands the two-story Hall of Tathagata, also called the Wanfo Pavilion (the Pavilion of the Ten-Thousand Buddhas), which is five bays wide with a hip roof covered with glazed black tiles. There are more than 9,000 little niches installed in the walls (most of them have been lost), and therefore the entablature of its upper eave was inscribed with the characters "Wan Fo Ge" (the Pavilion of the Ten-Thousand Buddhas). The ceiling of the bright room inside the pavilion was decorated with eight exquisitely carved sunk panels engraved with cloud dragons. However in the 1930s, they were stolen and sold by some monks, and they are now preserved in the Nelson Museum in the United States. In the front of the Hall was built a platform with outstretched walls in the east, south and west, which were all buried by the

▲ The Wanfo Pavilion

▲ Statues of Ten-thousand Buddhas in the Wanfo Pavilion

soil. Originally named the Jile Hall (the Bliss Hall), the Hall of Tathagata is now called the Hall of Great Mercy which is three bays wide with a gable-and-hip roof supported by a main ridge and covered with glazed black tiles.

In the east of the Hall of Great Mercy was situated the Abbot Hall whose original structures were non-extant now and its structures built later were used for other purposes.

▲ Beijing Local Music Show in the Zhihua Temple

# The Hongluo Temple

The Hongluo Temple (The Red Snail Temple) is located at the Hongluo Mountain in the north of Huairou District in Beijing. It is the largest Buddhist temple in northern China where Buddhists are practicing the Buddhism school of Pure Land. It was designated by Beijing Municipal Government as a major historical and cultural site to be protected in 1990.

The Hongluo Temple was first constructed in the 4th year (348) during Emperor Yonghe's reign in the Eastern Jin Dynasty and was named the "Daming Temple". It was renamed the "Huguo Zifu Buddhist Temple" during the years in Emperor Zhengtong's reign (1436-1449) in the Ming Dynasty. There is a "Pearl Spring" at the foot of the mountain and according to the legend, there were two large dark red snails living in the deep water of the spring and they spat red luminous flame thus the mountain was named "the Red Snail Mountain" and the temple was called "The Red Snail Temple".

The south-facing Hongluo Temple covers an area of more than 100 *mu* and is divided into three architecture complexes. The main structures in the central courtyard are as follows: Three main gate halls among which the middle one is the

▲ The Main Gate

beamless single-eaved main gate with a gable-and-hip roof. Backwardly erects the Hall of the Heavenly Kings that is five bays wide with a flush gable roof supported by a main ridge and covered with round tiles. The Mahavira Hall follows closely and was built on a base and five bays wide with a single-eave gable-and-hip roof. In front of the hall was built a platform. The main hall is flanked with side halls on its east and west sides. The last hall in the middle courtyard is the Three Kegon Hall that is five bays wide with a flush gable roof supported by a main ridge and covered with round tiles. The west side yards serves as the monks' residence. The west side yard includes the Dressing House and the living room. In addition, the temple also has kitchens and residences for the personnel regularly doing certain odd jobs. Outside the west wall stand the pagoda complexes which are the monks' tomb pagodas.

▲ The Mahavira Hall

▲ Interior View of the Mahavira Hall

▲ The Three Kegon Hall

Buddhist Temples

115

# The Cheng'en Temple

Situated in the north of the Moshikou Street Dongkou in Shijingshan District of Beijing. It is distinguished from the other temples for its watchtowers around it. It was designated by the State Council as the major historical and cultural site under state protection in 2006.

The Cheng'en Temple was first founded during the years in Emperor Wude's reign in the Tang Dynasty and was later rebuilt on its original site by Eunuch Wen Xiang in the 5th year (1510) and was completed in the 8th year (1513) in Emperor Zhengde's reign in the Ming Dynasty. It went through two restorations in the 22nd year (1757) in Emperor Qianlong's reign and in the 23rd year (1843) in Emperor Daoguang's reign in the Qing Dynasty.

The south-facing Cheng'en Temple has four courtyards fenced by walls. It covers an area of six thousand square meters. Its three-bay main gate has a

▲ Yard Wall and the Watchtower

▲ The Main Gate

Buddhist Temples

117

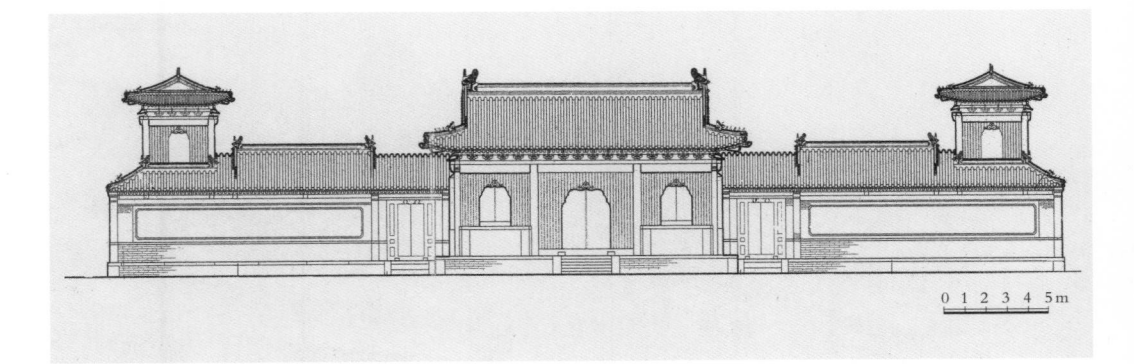

▲ Elevation of the Hall of the Heavenly Kings

gable-and-hip roof covered with round tiles. On the stone board above the white marble arched doorway were inscribed characters "Chi Jian Cheng'en Chan Si" (Construction of the Cheng'en Temple under Imperial Command). A side door was installed on both sides of the gate. To the north of the front gate erects the Hall of the Heavenly Kings which is three bays wide with a gable-and-hip roof covered with round tiles. Under the eaves is a bracket set whose wooden framework was

▲ Wall Painting in the Hall of the Heavenly Kings (No.1)

decorated with colored tangent circle patterns. The wall was painted with dragon-style frescos of the Ming Dynasty. There are three corner rooms on the two sides of the hall. Buildings were constructed around the corner where the Bell Tower stands in the east and the Drum Tower stands in the west. Architectural structure of this type is rare among the temples built in the Ming Dynasty. To the north of the Hall of the Heavenly Kings stands the five-bay Mahavira Hall that is nine purlins deep with a gable-and-hip roof covered with round tiles. There are verandas at the front and back of the hall and its wooden structure was decorated with colored tangent circle patterns. A delicate 2.5-meter-high copper statue of Sakyamuni was molded and seated on the Lotus Throne before. Three steles stand in the front of the hall. Set up in the years of Emperor Zhengde, the first stele was inscribed with the characters "Chi Ci Cheng En Si Bei" (Construction of the Cheng'en Temple under Imperial Command); the second stele was renovated in Emperor Qianlong's reign; the third one was inscribed with the characters "Renovation of the Cheng'en Temple" set up during the years of Emperor Daoguang's reign. On the east and west side of the Mahavira Hall stands three side halls with a flush gable roof covered with round tiles. Under the eaves is a Maye-styled bracket set with a slanting masonry structure whose wooden structure was decorated with colored tangent circle patterns. In the

▲ Wall Painting in the Hall of the Heavenly Kings (No.2)

north of the side hall are situated seven side rooms with a front veranda. To the north of the Mahavira Hall lies the five-bay Dhamma Hall that is seven purlins deep with a flush gable roof covered with round tiles and with two verandas in the front and at the back that were decorated with colored tangent circle patterns. The east and west side of the Dhamma Hall were flanked with four sides rooms. Three side halls are on both the left and right sides between the Mahavira Hall and the Dhamma Hall.

Watchtowers were built on the four corners of the temple. Inside the temple, tunnels were built to connect the basement and the watchtowers, which were rarely seen in the temples in the Ming and Qing Dynasties. It was said that the Cheng'en Temple was built by the famous eunuch Liu Jin in the Ming Dynasty. Cheng'en Temple became his base camp for military training and revolt because he was unsatisfied with the addressing of "Jiuqiansui" (It means "nine thousand years old", indicating he was only inferior to the Emperor who was addressed "ten thousand years old"). It was also said that Cheng'en Temple was the important peripheral stronghold for the Ming's spy agency Dongchang and Xichang and it also functioned as an imperial intelligence organization. Analyzed from the framework of the temple, this temple should have assumed military functions in a sense though the legends were not supported by historical records.

▲ Underground Passage

▲ The Mahavira Hall

# The Guanghua Temple

Located at No. 31Houhaiyaer Lane in the Xicheng District of Beijing, the Guanghua Temple is a famous Buddhist temple in Beijing. It was designated by the Beijing Municipal Government as a major historical and cultural site to be protected in 1984.

The Guanghua Temple was first built in the Yuan Dynasty and got renovated during the years in the reign of Emperor Tianshun and Chenghua. Owing to the financial support of Eunuch Su Cheng in the palace, the renovated Guanghua Temple is very large. By the 27th year (1599) in the reign of the Ming Emperor Wanli, Guanghua Temple had become a Pure Land to spread Buddhism. The abbot of the temple led all the monks to hold an unprecedented Amitabha Dharma assembly. It was renovated again in the 20th year of Emperor Guangxu's reign (1894) in the Qing Dynasty. Zhang Zhidong put all the books he stored in the temple in 1908 and presented a request to the Qing Government for setting up the Capital Library. It was permitted in the next year and the Qing Dynasty assigned

▲ The Main Gate

▲ The Five-bay Mahavira Hall

Miu Quansun to take charge of the foundation of the library. After the foundation of the Republic of China, The Minister of Education, Cai Yuanpei, appointed Jiang Han as the head of the library and it received readers in the next year. As the chief of the first section of the Social Education Department of the Ministry of Education, Lu Xun frequented here. The library was moved to another place and the Guanghua Temple continued to serve as a Buddhist temple. Artist Pu Xinyu, a lay Buddhist, contributed to renovating the temple once again in 1938. In 1939, the Guanghua Temple founded the Guanghua Buddhist Academy that recruited dozens of monks as students and invited Buddhist scholars such as Zhou Shujia, Wei Shanchen, Xiu Ming, Haicen, Pu Xinyu etc. to teach Buddhism here. Guanghua Primary School was founded later and it forgave tuitions from the students and provided books and stationery for the students from poor families. The Education Bureau of Beijing took it over until 1952. Master Xuyun visited Beijing and lived in the Guanghua Temple in September 1952. Many Buddhists all came here to pay their respect to the master of Buddhism and the senior monk, making the quiet temple once swarmed with visitors.

The Guanghua Temple faces south and its structures laid out in the east, central and west courtyards seem grand and solemn. This temple houses 329 halls and rooms. The main buildings are located in the central courtyard. Along the central axis are situated the main halls that can be listed as follows: the screen wall, the three-bay main gate that had a gable-and-hip roof and a lintel inscribed with four big character "Chi Ci Guang Hua Si" (The Imperial Mandate to Name it the Guanghua Temple) written with gold powders, the three-bay Hall of Heavenly Kings with a hip roof, the five-bay Mahavira Hall with double eaves and a gable-and-hip roof, the two-story Pavilion for Storing Buddhist Scriptures with a flush gable roof where are laid out the bilaterally symmetrical Bell Tower, Drum Tower, the Hall of Garan, the Hall of Patriarch, etc.

The east courtyard included the Ordination Altar, the Abstinence Hall, the Precept Learning Hall in the past.

The courtyards in the west are surrounded by galleries and adjoined by monks' residences, which forms a big courtyard that houses many small courtyards, namely the typically structural characteristic of "courtyard in courtyard". The temple is filled with luxuriant ancient pines and cypresses, fragrant flowers and winding paths.

In addition, the Guanghua Temple stored 1716 items of national cultural heritages, among which are the 1087 books, 282 calligraphies and paintings, 298 rubbing from a stone inscription, and 49 items of other cultural heritages. Some of the items in the collection are treasures of cultural heritages such as the *Buddha Vatamsaka-mahavaipulya Sutra* engraved during the years in the Ming Emperor Yongle's reign, the *Vajracchedika Sutra* transcribed by the Qing Emperor Yongzheng, and many calligraphies and paintings created by the celebrities in the Ming and Qing Dynasties. The temple collects four precious volumes of the *Tripitak*, totaling 2761 cases.

▲ Statues of Buddha in the Five-bay Mahavira Hall

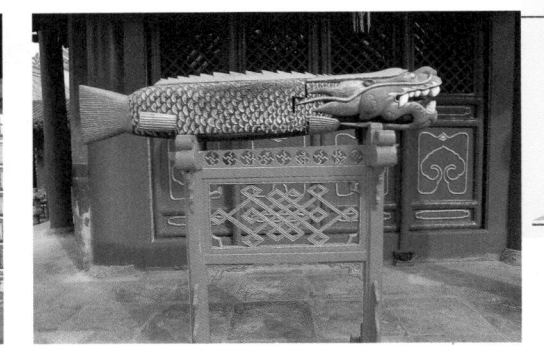
▲ Wooden Fish

▲ The Ksitigarbha Hall

▲ The Dhamma Hall

# The Dahui Temple

Located on Dahui Temple Road of Weigongcun in Haidian District, Beijing. It was designated by the State Council as the major historical and cultural site under state protection in 2001.

The Dahui Temple was first built by Eunuch Zhang Xiong, who worked in the Rites Supervising Office, in the 8th year (1513) in the Ming Emperor Zhengde's reign. The Hall of Great Mercy was the only structure existent in the temple.

The Hall of Great Mercy is five bays wide and three bays deep with a double-eaved hip roof covered with round gray tiles. The windows used for lighting were decorated with water chestnut flowers. Colorful small Buddhas were put on the square columns in the temple. A five-*zhang* (one *zhang* = 3.33 meters ) tall copper Avalokitesvara Bodhisattva with a Thousand Arms and Eyes used to be seated in the outer room hence the temple was also called the Huge Buddha Temple. The copper statue was destroyed in the War of Resistance against Japan. The two disciples

▲ The Hall of Great Mercy

carved by color wood were molded later. 28 color clay statues of the Heavenly Gods made in the Ming Dynasty were seated on the two gable walls and before the wall of the rear eave in the temple. The 28 statues of the Heavenly Gods are the Guardian Deities to the Buddha. According to the different personalities of the Heavenly Gods, the outstanding sculpture artists molded different expressions on the statues and decorated them with clothes and various colors, which made the statues vivid and imposing and become the sculpture masterpieces in the Ming Dynasty existent today. On the wall at the back of the Heavenly Gods was painted a set of large-scale colorful frescos that told a story about a life-long practioner of good deeds who received an after life. The theme of the story was unique and the fresco was bright in color with life-size and meticulously carved characters. It is really rare for a temple in Beijing to preserve the structure, engravings and frescos of the Ming Dynasty at the same time.

▲ Color Sculpture in the Hall of Great Mercy in the Ming Dynasty (No.1)

▲ Color Sculpture in the Hall of Great Mercy in the Ming Dynasty (No.2)

▲ Wall Painting in the Hall of Great Mercy

Buddhist Temples

# The Mohe Nunnery

The Mohe Nunnery is located at the Balizhuang Primary School in Sijiqing County, Haidian District of Beijing. It is one of the important temple buildings in western Beijing and especially noted for storing the Vajra Sutra steles in the Ming Dynasty. It was designated by Beijing Municipal Government as the major historical and cultural site to be protected in 1995.

The Mohe Nunnery was first built in the 25th year during the reign of Emperor Jiajing in the Ming Dynasty (1546). Eunuch Zhao Zheng raised funds and constructed this nunnery. According to the historical record, "There are two steles in the front of the Mohe Nunnery. The inscriptions on the left and right steles were written respectively by the minister Fei Cai from Qianshan and the minister Sun Cheng'en from Huating. The two steles were both set up in the 27th year during the reign of Emperor Jiajing." The Mohe Nunnery is located in the area of Balizhuang where the environment is elegant, and luxuriant pines and cypresses are growing. The beautiful environment attracted numerous tourists and also provided a good place for officials, poets and literary men to go for spring outings, appreciate peach blossoms and lilacs and compose poems in the Ming and Qing Dynasties, which made it a famous temple in western Beijing at that time. It was used as a school in the Republic of China Period and continues the same function now.

The south-facing Mohe Nunnery was divided into three parts. The hall area in the middle of the nunnery had a complete preservation. The main hall, the fourth courtyard in the Dhamma Hall area in the east of the nunnery, is the only building existent now. The main house, the fourth courtyard in the monk's residential area in the west is the only structure extant. The tomb of Zhao Zheng was once located in the nunnery but has been destroyed and buried now. On each corner of the surrounding areas near the enclosure wall of the nunnery was built a watchtower-shaped turret by tiger-skin ashlars. Only three of them are existent now and the turret in the southwest corner was destroyed.

The south-facing main gate is situated in the front of the mid-way courtyard and is three bays wide with a gable-and-hip roof covered with round tiles. The

▲ The Watchtower

▲ The Main Gate

stone lintel under the eaves was inscribed with three characters "Mo He An" (the Mohe Nunnery). In the front eave were opened with an arched doorway and two red wooden slab doors inlaid with stones. The back eave was in a shape of an ice tray. On the beam was painted with two colored half-tangent circle patterns. A wall door was inlaid in the east and west side of the main gate. At the back of the main gate stands the three bays wide Hall of the Heavenly Kings with a gable-and-hip roof covered with round gray tiles. Under the eave was installed a Maye-styled bracket set. The outer room under the front eave was designed with an arched doorway and the side room was opened an arched stone window. They were all covered with dripping tiles and decorated with bricks. The Hall of the Heavenly Kings was built on a brick base and a pair of stones for mounting the horse carved with patterns of auspicious clouds and galloping horses stands in the forecourt of the hall. The hall also has a pair of square-column-shaped flagpole stones carved with lotus flower patterns. The three-bay Mahavira Hall has a gable-and-hip roof covered with round gray tiles. Under the eave was constructed a bracket set whose architrave was painted with big-dotted half-tangent circle patterns. Behind the back eave was constructed a platform under which were buried the steles in the temple. The roof of the main hall was covered with three glazed yellow tiles that were granted by the emperor when the temple was constructed. Three side halls stand on the east and west of the hall with a hip roof supported by a main ridge and covered with round tiles. Under the eave stands a Maye-styled bracket set, which was decorated with gold thread big-dotted golden tangent circle pattern and golden-dragon lintels. The five back halls are six purlins deep with a flush gable roof covered with tiles and have a gallery stretching out of the hall. The architrave of the hall was painted with a half-tangent circle pattern and the lintel of the hall was painted with a golden dragon. Under the front eave stand four-checkered partitions and the side room has three sill wall windows that all have a window with three sashes and six grills. The five halls were built on a block stone base.

The south-facing Hall of King Kong in the east part was three bays wide and six purlins deep with a gallery stretching out of the hall and a flush gable roof supported by a raised ridge and covered with round tiles. The architrave of the hall was painted with a half-tangent colored circle pattern. The lintel of the hall was painted with golden dragons. The floor in the hall was paved with quadrels and the wall was inlaid with great *Vajra Sutra* engraving stone blocks. The blocks were carved by 60 cubic meters stones and were the rare cultural heritage for studying the Chinese calligraphy and classical Buddhist scriptures. There was a tiger-skin-colored stone wall around the courtyard on the east side.

The fourth courtyard in the west part of the temple has three main rooms and

▲ The Hall of the Heavenly Kings

▲ The Mahavira Hall

▲ The Hall of King Kong

is south-facing with a flush gable roof covered with round tiles and supported by a raised ridge. The architrave of the hall was painted with tangent colored circle patterns.

▲ *Vajra Sutra* Engraved on Stone Blocks in the Hall of King Kong

# The Changchun Temple

The Changchun Temple is located at No. 7, 9 and 11 Changchun Street in Xuanwu District. The temple is among one of the few perfectly preserved grand temples in Beijing. It was considered as the "Largest Temple" in the Ming Dynasty and was designated by the Beijing Municipal Government as a major historical and cultural site to be protected in 2001.

The Changchun Temple was first constructed in the 20th year (1592) of Emperor Wanli's reign in the Ming Dynasty. By imperial command of Empress Dowager Xiaoding, it was built for Shui Zhai (his monastic name), a Master of Buddhism named Ming Yang as his residence. Emperor Shen, Zhu Yijun, named it "the Changchun Temple" meaning longevity and good health of the Empress Dowager. At the beginning of the Qing Dynasty, the famous scholar Gong Dingzi contributed to construction of a pavilion in the Changchun Temple and named it the Miaoguang Pavilion (The Wonderful Light Pavilion) for the scholars to stand high, look afar and create poems. The Changchun Temple was destroyed by earthquake in the 18th year (1679) of Kangxi's reign and was renovated later by Feng Gongpu from Yidu. Qian Jinxi from Jingzhao established the Wanping Free School in the 39th year (1700) of Kangxi's reign. The Changchun Temple underwent a second renovation during the years of Qianlong's reign. During the years of Kangxi's reign, the Miaoguang Pavilion collapsed and was rebuilt during the years of Jiajing's reign and renamed the Zhe Temple. After Li Dazhao was murdered in April 1927, his corpse was placed in the Changchun Temple and was later transferred to a southern room of the Zhe Temple, which has been pulled down now.

The main architectures of the east-facing Changchun Temple gather on an east to west axis and consist of four halls. The one-bay main gate has a hip-and-gable roof covered with round tiles and the horizontal stone tablet was engraved with the characters "Chi Jian Chang Chun Si" (Construction of the Changchun Temple by Imperial Command). The three-bay main hall in the first courtyard is five purlins deep with a flush gable roof supported by a main ridge and covered with round tiles. The three-bay main hall in the second courtyard is nine purlins deep with a corridor in the front of it. In the south and north behind the main hall stand three side halls, all with a flush gable roof covered with glazed yellow tiles. The third courtyard houses the Pavilion for Storing Buddhist Scriptures that is five bays wide and seven purlins deep with a corridor in the front of it. The fourth five-bay courtyard has two connected roofs. On the two sides of the main structures were built side rooms. On the northern side of the central courtyard were built the northern side hall and several small architecture groups. The Changchun Temple serves as the Beijing Xuannan Culture Museum open to the public now.

▲ The Main Gate

▲ Halls in the Changchun Temple

▲ The Main Hall in the Second Courtyard

Buddhist Temples

133

# The Juesheng Temple
# (The Big Bell Temple)

The Juesheng Temple (the Temple of Awakening) is located on the north side of the North Third Ring Road in Haidian District, Beijing. It was distinguished for storing a big bell moulded in the Ming Dynasty so it was also called "the Big Bell Temple". The temple was designated by the State Council as the major historical and cultural site under state protection in 1996.

The Juesheng Temple was first built in the 11th year (1733) and completed in the winter in the 12th year (1734) under the reign of the Qing Emperor Yongzheng. According to the record on a stone board, "There was a beautiful farmland in Zengjiazhuang Village outside Xizhimen in the city. The area grew luxuriant trees and was far from the city center and surrounded by green hills. It was a superbly quiet place for followers of Buddhism. A temple should be constructed there." The temple served as the venue for the monks and devotees to the Buddha to pray. Emperor Qianlong went to the temple to pray for rain in the 52nd year (1787) during his reign after Beijing suffered a severe draught. The imperial praying-for-rain services at the Big Bell Temple lasted until the end of the Qing Dynasty.

The south-facing Big Bell Temple has five courtyards. Standing from south to north were the major buildings like the Screen Wall (destroyed), the main gate, the Hall of the Heavenly Kings, the Hall of Mahavira, the Rear Hall, the Pavilion for

▲ The Main Gate

▲ The Mahavira Hall

Storing Buddhist Scriptures, the Great Bell Building and the east and west flanking halls. In addition, there are Bell and Drum Towers and six side halls distributed on the two sides of the temple. The main gate stands on the most southerly part of the temple and is three bays wide and five purlins deep with a gable-and-hip roof suspended by a main ridge and covered with round tiles. Under the eaves is a bracket archway with a slanting masonry structure whose wooden structure was decorated with colored tangent circle patterns. The outer room has six pavilion roofs and the side room has five pavilion roofs with a wooden structure painted with ink markers. The roofs were decorated with colored golden tangent circle patterns. The crouching board above the stone arched doorway were inscribed the characters "Chi Jian Jue Sheng Si" (Construction of the Juesheng Temple by Imperial Command). The side room has arched stone windows. On both sides of the main gate stands a side gate flanked with a screen wall to hide the gates from the mountains. The four angles of the screen wall were embedded with a patterned brick carving and had a high molding-decorated base. A pair of stone lions stand on both the left and right sides of the gate. The Bell and Drum Towers are located in the first courtyard inside the main gate. The framework of the flanking structures on their two sides is identical. They all have a gable-and-hip roof and covered with round tiles. Under the eaves is a bracket archway with a lean masonry structure. The upper layer of its wooden structure was painted with big-dotted ink-line golden tangent circle patterns. The light barrier on the four sides of the upper layer was designed with four curling windows. A stone arched doorway was installed under the front eave of the lower layer.

The Hall of the Heavenly Kings is the main hall in the first courtyard and is three bays wide and five purlins deep with a hip roof suspended by a main ridge and covered with round tiles. Terrific small bells were positioned on the roof whose wooden structure was painted with big-dotted ink-line golden tangent circle patterns. The light barrier on the front eave was designed with a curling door and window. The outer room at the back eave had four five-checkered fan doors. The Mahavira Hall is the main hall in the second courtyard and is five bays wide with a hip roof suspended by a main ridge and covered with round tiles. In front of the Mahavira Hall stand three round-ridge-roofed verandas whose wooden structure was painted with big-dotted ink-line golden tangent circle patterns. Three sides of the platform in front of the hall have outstretched walls. There is a five-step stone footstep

paved with slanting boulder strips leading to the platform. On both the eastern and western sides inside the courtyard stand five halls with a veranda stretched out and hip roofs supported with a main ridge and covered with round tiles. Their wooden structures were painted with small-dotted ink-line golden tangent circle patterns. The southern side of all the side halls was flanked with a side room. The Rear Hall is the main hall in the third courtyard and is five bays wide and seven purlins deep with a hip roof supported by a main ridge and covered with round tiles. Terrific small bells were placed on the roof whose wooden structure was painted with big-dotted ink-line golden tangent circle patterns. A five-step stone footstep paved with two slanting boulder strips is standing in the front and at the back of the hall. On both the east and west sides of the hall were laid out 15 side rooms with a hip roof supported by a main ridge and covered with round tiles. The timber structures of the side rooms were painted with tangent circle patterns. On the north of the fourth courtyard stands the two-story Pavilion for Storing Buddhist Scriptures The seven-bay hall is seven purlins deep with a veranda in the front and at the back. This hall has a hip roof supported by a main ridge and covered with round tiles. Terrific small bells were placed on the roof whose timber structure was painted with big-dotted golden tangent circle patterns. The front eaves on the second floor were decorated with railings covered with lateral frames. The stairs on both sides of the gable walls lead to the second story of the hall. The courtyard houses five side halls in the east and west with a veranda in the front and at the back. The side halls all have a hip roof supported by a main ridge and covered with round tiles. Terrific small bells were placed on the roofs whose timber structural framing was painted with small-dotted ink-line golden tangent circle patterns. All the side halls were installed with partitions and windows. There are seven side rooms on both the east and west side.

Located in the last courtyard of the temple, the 20 meters high Big Bell Tower is a unique core structure in the temple. With a seven-step stone footstep paved with two slanting boulder strips on its two sides, it erects on the platform based on the bluestone base and fenced by white marble stone boards and baluster columns. The round top and square base symbolize that "the heaven is round and the earth is square." The upper layer of the tower has a round pavilion roof covered with round tiles. Under the eaves is installed a bracket set with a slanting masonry structure. The round tower is divided into 12 rooms by 12 columns and each room has a bracket set with a quadrangle roof. Under the eaves are four square windows with

▲ Characters Cast on the Yongle Bell

▲ The Big Bell Tower

▲ The Ceiling of the Big Bell Tower

▲ Knot on the Top of the Yongle Bell

boards hanging underneath. The three-bay lower part of the Bell Tower has a bright room with a six-checkered fan door above which hangs a crouching board engraved with the characters "Hua Yan Jue Hai". A five-step stone footstep paved with two slanting boulder strips stands in the front of the door and the side rooms were designed with four-checkered fan doors. The bell hangs in the tower where a stair can be used for climbing up and down. On the bluestone base is made an octagonal "pond for diffusing sound" which is 70 centimeters deep and four meters in diameter. The inlet of the pond is one meter away from the bell end. When the bell rings, the sound produces echoes. Molded in the years in Emperor Yongle's reign in the Ming Dynasty, the Yongle Bell is 6.75 meters high, 3.7 meters in diameter and weighs 46.5 tons. The edge of the bell is 18.5 centimeters thick. The bell is bright without any cracks. Besides, it was cast entirely with Buddhist sutras totaling 230,184 characters. It is an unequalled Buddhist bell with exquisite casting craft and is the treasure of the culture of Buddhism. The flanking halls are standing on the east and west of the Big Bell Tower.

Buddhist Temples

▲ The Yongle Bell

# The Lingyue Temple

The Lingyue Temple is located at the Baitie Mountain five kilometers away from the north of Zhaitang Town in Mentougou District of Beijing. It was designated by Beijing Municipal Government as a historical and cultural site to be protected in 2003.

The Lingyue Temple was first constructed during the years in Emperor Zhenguan's reign in the Tang Dynasty. It was rebuilt in the Liao Dynasty and was called "Baitie Mountain Temple" and renamed "Lingyue Temple" in the Jin Dynasty. It was renovated in the 30th year (1293) during Emperor Zhiyuan's reign in the Yuan Dynasty and during years in Emperor Zhizheng's reign in the Yuan Dynasty. It underwent two renovations in the 22nd year (1683) during Emperor Kangxi's reign and in the 11th year (1733) during Emperor Yongzheng's reign in the Qing Dynasty.

The south-facing temple is located on the terrace before the main peak of the Baitie Mountain. On its central axis stand the main gate, the Hall of the Heavenly Kings and the Mahavira Hall. The ruins of the Bell and Drum Towers are located on the two sides of the main gate. The three-bay Hall of the Heavenly Kings has an overhanging gable roof. It was renovated by successive dynasties but still preserves the architectural style of the Yuan Dynasty. In the hall are enshrined the Four Heavenly Kings, Skanda, and the statue of the Ambassador Buddha. The five-bay Mahavira Hall has a single-eaved hip roof. Under its eave is a bracket set in a Wucai style. The eye wall on the arch was painted with colored figures of Buddha. One Buddha and two Bodhisattvas were enshrined in the hall. They were carved with willow and were four meters high. The statue of Sakyamuni was seated on the lotus-shaped seat and the statues of Ananda and Kasyapa were sitting on both sides of Sakyamuni. The lifelike carved statues were pulled down in 1954. What still exists in the temple are the stele inscribed with the "Prose of Renovating Lingyue Temple" and the stele engraved with the "Prose of Renovating Lingyue Buddhist Temple". The previous was set up in the 30th year (1293) in Emperor Zhiyuan's reign in the Yuan Dynasty and the latter was set up in the 22nd year (1683) in Emperor Kangxi's reign in the Qing Dynasty.

▲ Panoramic View of the Lingyue Temple

▲ Frame of Beams of the Mahavira Hall

▲ Upturned Roof-corner of the Mahavira Hall

▲ Exterior View of the Temple

▲ The Hall of the Heavenly Kings

▲ Bracket in the Hall of the Heavenly Kings

▲ The Mahavira Hall

Buddhist Temples

141

# The Main Hall of the Lingyan Temple

Located in Qijiazhuang Village, Qingshui County in Mentougou District, Beijing, The Great Lingyan Temple is one of the few Buddhist temples where the architectural style of the Yuan Dynasty is preserved. It was designated by Beijing Municipal Government as a historical and cultural site to be protected in 1995.

The Lingyan Temple was first constructed during the years in Emperor Wude's reign in the Tang Dynasty and was rebuilt in the 1st year during Emperor Zhizheng's reign in the Yuan Dynasty. It was renovated in the 22nd year (1486) during Emperor Chenghua's reign in the Ming Dynasty and in the 6th year (1527) during Emperor Jiajing's reign in the Qing Dynasty. The prosperity of the temple was on the decline after the Qing Dynasty.

The south-facing Lingyan Temple originally had a main gate, Bell and Drum

▲ The Main Hall

▲ Pagoda Ridge Ornament on the Main Hall

Towers, the Prince Hall, the Garan Hall, the Mahavira Hall and its side halls, which were destroyed by the Japanese Army during the War of Resistance against Japan. What is existent now is the Mahavira Hall that was built on a stone base. The three-bay hall covers an area of 118.5 square meters. The inner structure of the hall used fewer columns and the bracket set under the eaves was supported by accessory structures. The whole beam frame shows the architectural crafts in the Yuan Dynasty.

▲ Beast Sculpture on the Main Hall

Buddhist Temples

# The Relics of the Cross Temple

Located at the foot of Sanpen Mountain, Mao'er Peak, 1.5 kilometers northwest to Chechang Village, Zhoukoudian County, Fangshan District, Beijing, the Cross Temple is the only Nestorianism (a branch of Christianity) remains in Beijing. It was designated by the State Council as a major historical and cultural site under state protection in 2006.

Nestorianism was a branch of Christianity. Introduced to China in the Tang Dynasty, it was the first branch of Christianity introduced to China.

▲ Panoramic View of the Relics of the Cross Temple

According to records on the stele, the Cross Temple was first built in East Jin Dynasty as a Buddhist temple. It was renovated in the years under Emperor Yingli's reign in the Liao Dynasty and renamed the Chongsheng Temple. In the 18th year in the Yuan Dynasty (1358), it was renamed again the Cross Temple which became a temple of Nestorianism. In the Ming Dynasty, it was transformed to a Buddhist temple which gradually withered away. In the Republic of China Period, two blocks of stone on which were carved a crossing and ancient Syriac scripts. They are preserved in Nanjing Museum.

Facing south, the Cross Temple covers an area of 2,000 square meters. The major buildings were built on the central axis symmetrically. Almost all of the structures were destroyed. What left are the enclosing wall, wall remains of the east wing room, the old ginkgo tree and two stone steles. One stele inscribed "Sanpen Mountain Chongsheng Temple Records" was erected in the 8th year of Emperor Yingli in the Liao Dynasty. The other inscribed "Construction of the Cross Temple under Imperial Command" was erected in the 23th year under Emperor Zhizheng's reign in the Yuan Dynasty.

▲ The Stone Stele

▲ Base of the Stone Columns

Buddhist Temples

# The Tiewa Temple

The Tiewa Temple  (The Iron Tile Temple) is located at the government courtyard of the Hebei Town in Mentougou District. It was named the Iron Tile Temple for the roof of the main hall was covered with iron tiles. The temple was designated by Beijing Municipal Government as a historical and cultural site to be protected  in 2003.

The main buildings in the south-facing Iron Tile Temple are the main gate, the Iron Tile Hall and the east and west side halls. The main gate is three bays wide and the lintel of the door was inlaid with a board on which were written "Tie Wa Chan Lin" (The Iron Tile Buddhist Temple). The Iron Tile Hall is round and has a pavilion roof covered with iron tiles. The roof was divided into six sectors by six ridges that droop in six directions from above. The number of iron tiles covering the roof is 485 pieces and each of them is 0.31 meter long and 0.13 meter in diameter. On the surface of the tile are casting characters such as "Pusa Ding—Constructed in the 10th Year during Emperor Zhengde's Reign" and "Pusa Ding Iron Tile Temple on Wutai Mountain".  It can be inferred that the Iron Tile Temple is related to Wutai Mountain. The roof and the ridge of the Iron Tile Hall were also iron-made. About three thousand kilograms of iron was used for constructing this hall. Such a structure is rarely seen in Beijing.

▲ Spire of the Iron Tile Hall

▲ Stone Board on the Main Gate

▲ The Iron Tile Hall

Buddhist Temples

## Tibetan Buddhist Temples

## The Miaoying Temple

The Miaoying Temple is located at No. 171 Fuchengmennei Street in Xicheng District of Beijing. It is also called the White Pagoda Temple because the entire exterior of the pagoda in the temple is painted by white lime. It is the most ancient Lamaist temple in Beijing. The White Pagoda in the temple is one of the earliest overturned bowl shaped buildings in Beijing. It was designated by the State Council as a major historical and cultural site under state protection in 1961.

Originally named the Temple of the Emperor's Longevity and Peace (Da Shengshou Wan'an Si), the Miaoying Temple was firstly built in the 8th year (1271) of Emperor Zhiyuan's reign. It was destroyed by fire in the 28th year (1368) of Emperor Zhizheng's reign. In the 1st year (1457) of Emperor Tianshun's reign in the Ming Dynasty, the Miaoying temple was rebuilt and the White Pagoda was renovated. After that the temple was named "the Miaoying Temple" by imperial command. Most of the structures in the temple were constructed in the Qing Dynasty. The White Pagoda is the only relic built in the Yuan Dynasty, over 700

▲ The Main Gate

▲ The Hall of the Heavenly Kings

years ago.

The south-facing Miaoying Temple contains four sets of courtyards. Imitating the timber-structure, the brick-structured main gate is three bays wide and one bay deep with a gable-and-hip roof supported by a main ridge and covered with round tiles. The bright side hall of the main gate comprises three arched doorways. Above the arched doorway of the outer room hangs a stone board inscribed the characters "Chi Ci Miao Ying Chan Lin" (The Miaoying Buddhist Temple Named by Imperial Command). On each of the two sides of the main gate erects a splay screen wall. A room stands on each of the east and west side of the main gate.

On the north side of the first courtyard stands the Hall of the Heavenly Kings which is three bays wide and five purlins deep with a gable-and-hip roof covered with round tiles. Under the eave was designed a bracket set whose timber structure painted with gold thread big-dotted golden tangent circle patterns. The outer room of the Hall of the Heavenly Kings was opened with a gate that was connected by a four-step stone footstep paved with two inclined boulder strips on its two sides. The gate was decorated with four three-way-crossed six-checkered fan doors. The side room was opened with four three-way-crossed six-checkered sill wall windows. The Maitreya Buddha and Skanda Buddha are enshrined in the temple with the statues of the Four Heavenly Kings on its two sides. On the east and west side of the first courtyard stand the Bell and Drum Towers having the same building structure. The two-story towers have a gable-and-hip roof and covered with round tiles. Bracket arches were designed under the eaves of the towers. The bright room on the lower level had an arched doorway.

The Yizhu Xinjing Hall is the main hall in the second courtyard. The Yizhu Xinjing Hall is five bays wide and three bays deep with a hip roof and covered with round tiles. Under the eaves of the hall was designed a bracket set whose timber structure was painted with gold thread big-dotted golden tangent circle patterns. A wooden board inscribed with the characters "Yi Zhu Xin Jing" (The State of Mind of Yizhu) hung in the outer room of the hall. The bright and side rooms in the front facade were decorated with four three-way-crossed six-checkered fan doors. In the front of the hall was constructed a platform with two columns on which were inscribed Buddhist scriptures. The east, west and south walls all stretched out. In front of the Yizhu Xinjing Hall stand three east and west side halls which all have a flush gable roof supported by a main ridge and covered with round tiles. The courtyard consists of three east and west side halls. Behind the Hall of the Heavenly Kings is a side room in the corner linking the east side hall to the west one.

Behind the Yizhu Xinjing Hall stands the Qifo Hall (the Hall of Seven

▲ The Yizhu Xinjing Hall

Buddhas) and a brick-paved path connects the two structures. The Qifo Hall is five bays wide and four bays deep with a hip roof and covered with round tiles. The central section of the main ridge was decorated with lucky vases. Under the eave was designed a bracket set whose timber structure painted with gold thread big-dotted golden tangent circle patterns. The bright and side rooms in the front facade were decorated with four three-way-crossed six-checkered fan doors. The room for stacking firewood was decorated with four three-way-crossed six-checkered sill wall windows. Only the outer room in the back facade had four three-way-crossed six-checkered fan doors. Seven Buddha statues are enshrined in the temple. On the east and west side of the Yizhu Xinjing Hall and the Qifo Hall stand the side rooms that lead straightly to the front of the pagoda complex.

The pagoda complex is located at the back of the Qifo Hall. The main hall has a gable-and-hip roof and covered with round tiles. Under the eave hung a board on which were inscribed the characters "Chi Jian Shi Jia She Li Ling Tong Bao Ta" (Construction of the Sakyamuni Lingtong Stupa by Imperial Command). Behind the main gate

▲ Sunk Panel of the Yizhu Xinjing Hall

▲ Parnashavari in the Hall of Seven Buddhas

Buddhist Temples

151

stands the Hall of Six Spiritual Penetrations that is three bays wide and seven purlins deep with a gable-and-hip roof and covered with round tiles. Under the eave was designed a bracket set whose timber structure was painted with gold thread big-dotted golden tangent circle patterns. The outer room was adorned with four three-way-crossed six-checkered fan doors annexed with curtain rods. The side room was opened with a three-way-crossed six-checkered sill wall window. Next to the side room stands an ashlar Buddha niche. The Buddha of the Past, the Present and the Future is offered in the hall. The White Pagoda is located behind the Hall of Six Spiritual Penetrations. It rises to a height of 50.9 meters. The pagoda is composed of three major parts: the base, the principal story and steeple, forming a dog-eared square pedestal seen from its plan view. With its top tightened by seven iron hoops, the body of the pagoda is shaped like an inverted alms bowl supported by a round of lotus pedals. The steeple was decorated with a thirteen-tier circular ring with a small top and a big bottom, which supports a 9.7-meter diameter bronze canopy. What hangs around the edge of the canopy is 36 bronze curtains carved with the Chinese characters and the eight treasures in the Buddhism. A windbell hangs at the bottom of each curtain. A bronze gold-glided hollow inverted-alms-bowl style pagoda mounts on the top of the steeple.

Some buildings are still existent on the west axis. At the south end of the west

▲ The Hall of Seven Buddhas

▲ The Hall of Six Spiritual Penetrations  ▲ Corner Pavilion in the Pagoda Yard

axis shows a single-hall-and-single-door-styled festoon gate. The timber structure under the eave of the gate was decorated with Suzhou-styled paintings and two slab doors were inlaid under the eaves. Four screen doors are showing behind the slab doors. On the east and west side of the festoon gate were built four galleries and cloisters, which lead to the east and west side halls. The main hall in the courtyard is the Dajue Nengren Hall with a flush gable roof supported by a main ridge and covered with round tiles. In the front of the hall was a gallery whose eaves were painted with small-dotted ink-line golden tangent circle patterns. The outer room and side room in the front facade of the hall were all decorated with four three-way-crossed six-checkered fan doors. The outer room in the back facade of the hall was decorated with four three-way-crossed six-checkered fan doors and the side room has four three-way-crossed six-checkered sill wall windows. In the courtyard stand three side halls on both the east and west side with a flush gable roof supported by a main ridge and covered with round tiles. In front of the side halls was a gallery and its eaves were painted with small-dotted ink-line golden tangent circle patterns. The outer rooms of the side halls were decorated with four three-way-crossed six-checkered fan doors and the side rooms were designed with removable windows. The galleries on the east and west side of the hall are linked to the main hall which is seven bays wide and six purlins deep with a flush gable roof and a round ridge roof. In front of the main hall stands a veranda. With a gable-and-hip roof and a round-ridge roof both covered with round tiles, a veranda stretches out of the

0  2  4  6  8  10m

▲ Elevation of the White Pagoda

porch of the outer room. The veranda and the front aisle carry with inversed lintels, blossom buds and faldstool lintels. The eaves were painted with small-dotted ink-line golden tangent circle patterns. The outer rooms were decorated with four three-way-crossed six-checkered fan doors and the second, the third and the last rooms were inlaid with removable windows. Behind the main hall erects the south-facing two-story Pavilion for Storing Buddhist Scriptures. The nine-bay hall (five bays on the north side, two bays on each of the east and west side) has a flush gable roof supported by a main ridge and covered with round tiles. A veranda stretches out of the hall and its two-story eave was painted with small-dotted ink-line golden tangent circle patterns. The steps of the veranda were adorned with decorated brackets and breast board. The eaves on the first story were adorned with eave-hung fascias and the steps were decorated with brackets and faldstool lintels. On the north side of the Pavilion for Storing Buddhist Scriptures stands a small courtyard with three north rooms and three south rooms. The buildings have a gable flush roof covered with round tiles. Some decorations were made to the buildings later.

# The Yonghegong Lama Temple

Located at No. 12 Yonghegong Street, the Yonghegong Lama Temple is the largest Tibetan Buddhist temple which is well preserved in Beijing. It was designated by the State Council as a major historical and cultural site under state protection in 1961.

Yonghegong Lama Temple was constructed at the old site of the Ming Dynasty's interior court palace for eunuchs who had official tiles in the 33rd year of Emperor Kangxi's reign in the Qing Dynasty(1694). It originally served as Emperor Yongzheng's residence before Yongzhen succeeded to the throne. After he ascended the throne, this temple became his old residence and was not allowed to live in by others. He changed half of the temple into the Upper House of the Gelug and the other half into a temporary imperial palace for the imperial family. The temporary imperial palace was destroyed later by fire. The Upper House was transformed into a temporary imperial palace in the 3rd year of Yongzheng's reign (1725) and was named "the Yonghegong Temple". Because the coffin of Yongzheng was placed in the temple in the 13th year of Yongzheng's reign (1735), the glazed green tiles on the main halls were replaced by glazed yellow tiles. The image of Emperor Yongzheng was enshrined in the Yongyou Hall which was hence renamed "the Shenyu Hall". Afterwards, Yonghegong Lama Temple became a palace for the Qing emperors to worship their ancestors. But most of the halls became the venues for the Gelug Lamas to read scriptures and hold Buddhist ceremonies. The Yonghegong Temple was officially changed into a lamasery and became the national center of Lama Administration for the Qing Dynasty in the 9th year during Emperor Qianlong's reign (1744). Yonghegong Lama Temple selected the abbot Lamas in many important lamaseries by drawing lots. Then the selected abbot Lamas were sent to the lamaseries across the country. The Lama Printing Service Office that was in charge of the nationwide teaching affairs was also founded in the east garden of the Yonghegong Temple. The Lamas in the temple were up to more than 500. After the foundation of the People's Republic of China, Yonghegong Lama Temple underwent a complete renovation. It was officially open to the public as a venue

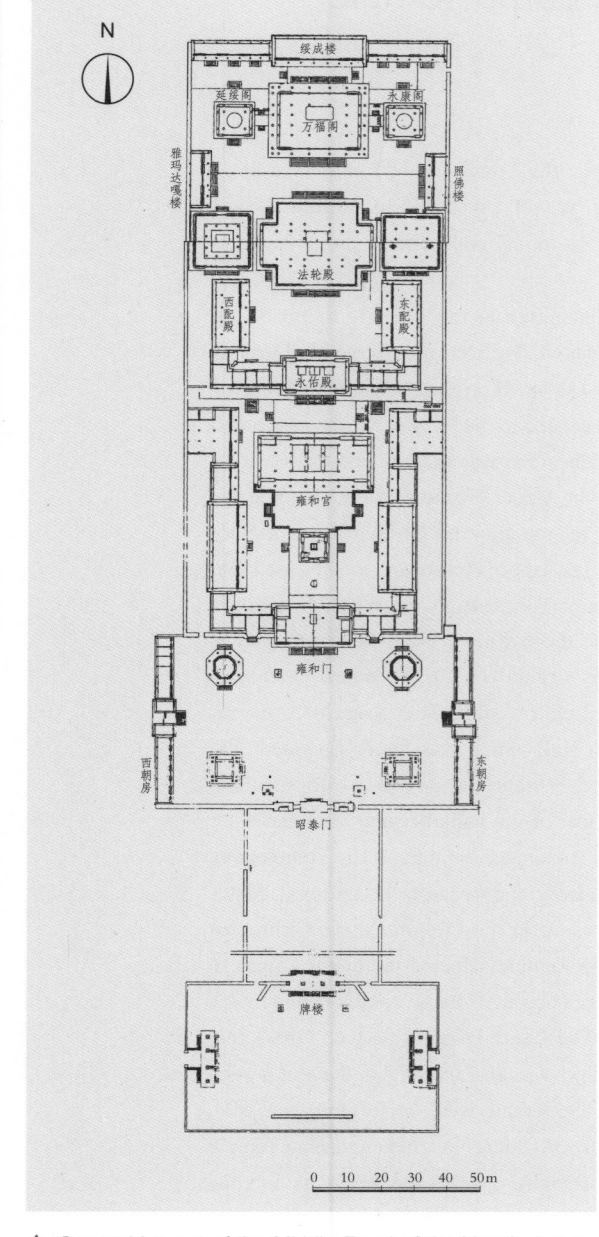

▲ General Layout of the Middle Road of the Yonghegong Lama Temple

for religious activities in 1981 and has become a distinguished tourist attraction.

The south-facing Yonghegong Lama Temple is divided into the central, east and west axis. The temple has features that combine the architectural features of the Han and Tibetan nationalities. The main structures are located on the central axis with five sets of courtyards. Transformed from an imperial palace, the temple is different in its front and back parts. The structures in the front part of the temple are spacious and palace-shaped. While those in the back part are of typical temple style and laid out orderly and interlaced. Its cultural heritage is matchless.

In front of the middle road in the temple stand three Pailous (decorated arches) on which were inscribed characters: "Huan Hai Zun Qin" and "Qun Sheng Ren Shou". A screen wall stands to the south of the Pailous. To the north of the Pailous was built an imperial road straightly leading to the main gate Zhaotaimen (Gate of Peace Declaration). The three-bay arch covered with glazed tiles was inlaid in the wall.

The first main hall is the Hall of the Heavenly Kings, which served as the main gate during the period when

▲ Bird's-eye View of the Yonghegong Lama Temple

Buddhist Temples

157

▲ The Zhaotai Gate

▲ The North Pailou

▲ Avenue behind the Pailou (The Imperial Road)

Buddhist Temples

▲ Gate of the Yonghegong Lama Temple

▲ Inscribed Board above the Gate of the Zhaotai Gate

▲ Interior View of the Gate of the Yonghegong Lama Temple

the temple served as the emperor's residence. It is five bays wide with a gable-and-hip roof. The statues of the Maitreya Buddha, the Four Heavenly Kings and Skanda are enshrined in the hall. There are two pavilions built over stone tablets and the Bell and Drum Towers in the courtyard. There are two side gates on the east and west side of the main gate and they were built based on imperial norms when Yonghegong Lama Temple served as the emperor's residence.

The second main hall is the Yonghegong Hall (the Hall of Harmony and Peace) which was the original Yin'an Hall during the period when the Yonghegong Lama Temple served as the emperor's residence and is also the main hall and center of the Yonghegong Lama Temple. The seven-bay-hall has a gable-and-hip roof. Under the roof hangs a board on which were written "Yong He Gong" in four languages: Han, Manchu, Mongolian and Tibetan. The statues enshrined in the hall are the Buddha of the Past, the Present and the Future (Medicine Guru Buddha, Tagathe Buddha, Amitabha), Avalokiteshvara, Maitreya and the Eighteen Arhats. There are four annexes and side halls in the east and west of the Yonghegong Hall, which are also called "the Four Study Halls", namely the Preaching Hall, the Vajrayana Hall, the Mathematics Hall and the Hall of the Medicine Guru Buddha where the

▲ The Yonghegong Lama Temple

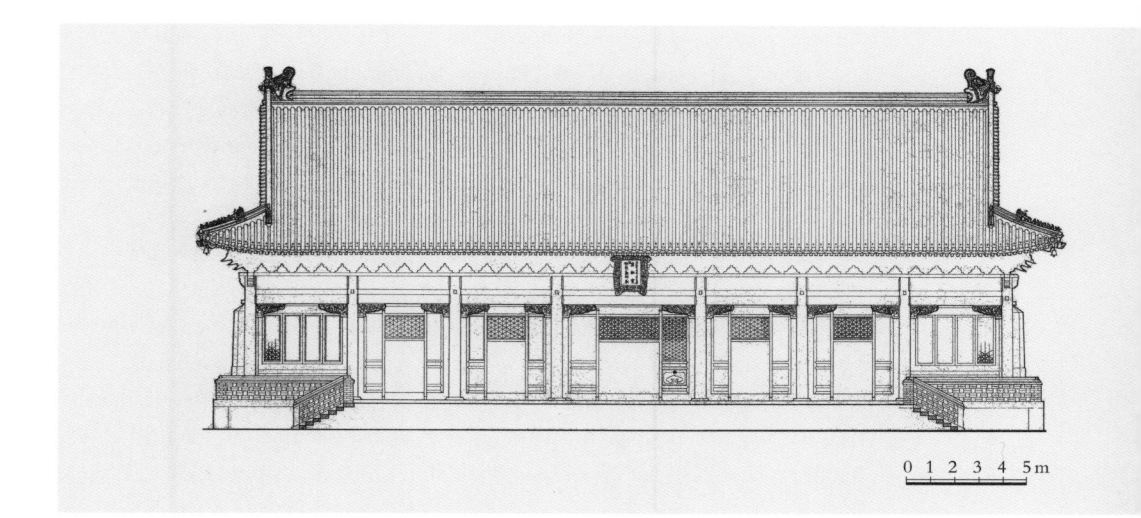

▲ Elevation of the Yonghegong Lama Temple

▲ Stele Pavilion of Lama Prose and Incense Burner (Lama Prose was Written by Emperor Qianlong on the Qing Government's Policy on Lamaism)

▲ Sunk Panel of the Yonghegong Lama Temple

monks study sutras of the Exoteric and Esoteric Buddhist Scriptures, math, calendar history and medicine. A bronze vessel cast in the 12th year of the Qianlong's reign (1747) stands in the Yonghegong Hall. The Qing emperors used it as an incense burner when they came here to worship. After the large incense burner is an imperial pavilion in which stands the tablet set up in the 57th year during Emperor Qianlong's reign (1792). On the tablet was inscribed the article Lama Prose in four languages: Han, Manchu, Mongolian and Tibetan. Emperor Qianlong wrote the article, which gave an account of the origin of the Lamaism and the Qing government's policy on Lamaism. Behind the stone tablet is a Mount Sumeru, a bronze sculpture made in the years of Wanli's reign in the Ming Dynasty.

The third main hall is the Yongyou Hall (the Hall of Everlasting Protection) which functioned as the inner gate during the period when Yonghegong served as the emperor's residence. It is five bays wide with a gable-and-hip roof. Three statues of Buddha are enshrined in the hall, with Amitayus Buddha in the center, Medicine Guru Buddha on the left and Roaring Lion Buddha on the right. The

▲ Interior View of the West Side Building of the Yonghegong Lama Temple (The Vajrayana Hall)

▲ East Side Building of the Yonghegong Lama Temple (The Preaching Hall)

▲ East Side Hall of the Yonghegong Lama Temple (The Mathematics Hall)

statues are 2.35 meters tall carved by sandalwood. On the west wall of the hall hangs a re-embroidered portrait of the Green Tara. It was said that the portrait was re-embroidered by Emperor Qianlong's mother Empress Dowager Xiaosheng. On the east wall hangs the portrait of the White Tara.

The fourth main hall was named the Falun Hall (the Hall of the Wheel of the Law)which served as the residence of the Emperor's wives during the period when the Yonghegong Temple served as the emperor's residence. The layout in the hall was similar to the Kunning Palace (The Palace of Earthly Tranquility) in the Forbidden City and the Qingning Palace (Hall of Peace and Tranquility) in the Forbidden City in Shenyang. The structures in the east part were used for dwelling place and those in the west part served as the Shamanism praying halls. It was transformed into the present cross-shaped layout during the years in Qianlong's reign. The seven-bay Falun Hall has a single-eaved gable-and-hip roof. There are five small cabinets in front of and behind the hall. On the roof of the hall erect five small attics designed by the Tibetan Buddhist tradition, which symbolize the five peaks of the Sumeru Mount. The tall hall functions as a place for all the Lamas in the temple to get together and read scriptures. It contains a huge 6.1 meters high bronze statue of Tsong Khapa, founder of the Gelug School of Buddhism. Behind the statue of Tsong Khapa comes the Five Hundred Arhats Hill which is one of the

▲ The Yongyou Hall

▲ Interior View of the Yongyou Hall

Buddhist Temples

165

"three unique treasures in the Yonghegong Temple". The 3.4 meters tall and 3.45 meters wide hill was exquisitely carved with sandalwood. On the hill are forests, valleys, pines and cypresses, pagodas, pavilions, caves, winding paths, bridges and flowing water. The statues of the 500 Arhats were made from five different metals — gold, silver, copper, iron and tin. The statues of the colorful and lifelike Arhats are an artwork whose material, modeling technique and carving skill are unique. Due to chaos caused by war, only 449 Arhats are left on the hill. This hall also stores 180 volumes of *Tripitaka*, 207 volumes of *On Tripitaka, the Medicine Guru Sutra* and the *True Buddha Liturgies* copied by Emperor Qianlong himself and some exquisite frescos. On the left of Falun Hall stands the Panchen Building (also called Medicine Guru Altar). It was named after the Sixth Panchen who visited Beijing to celebrate Emperor Qianlong's birthday in 1780 and expounded the Buddhist scriptures here. On the right side of Falun Hall comes the Ordination Altar on which were designed a folding screen and a throne. Putting on the court dress and wearing the royal scarlet hada (a piece of silk used as a greeting gift among the nationalities of Zang and Man) and the Buddhist crown, Emperor Qianlong used to sit quietly on the throne and practice meditation, and expound the Buddhist scriptures.

▲ The Falun Hall

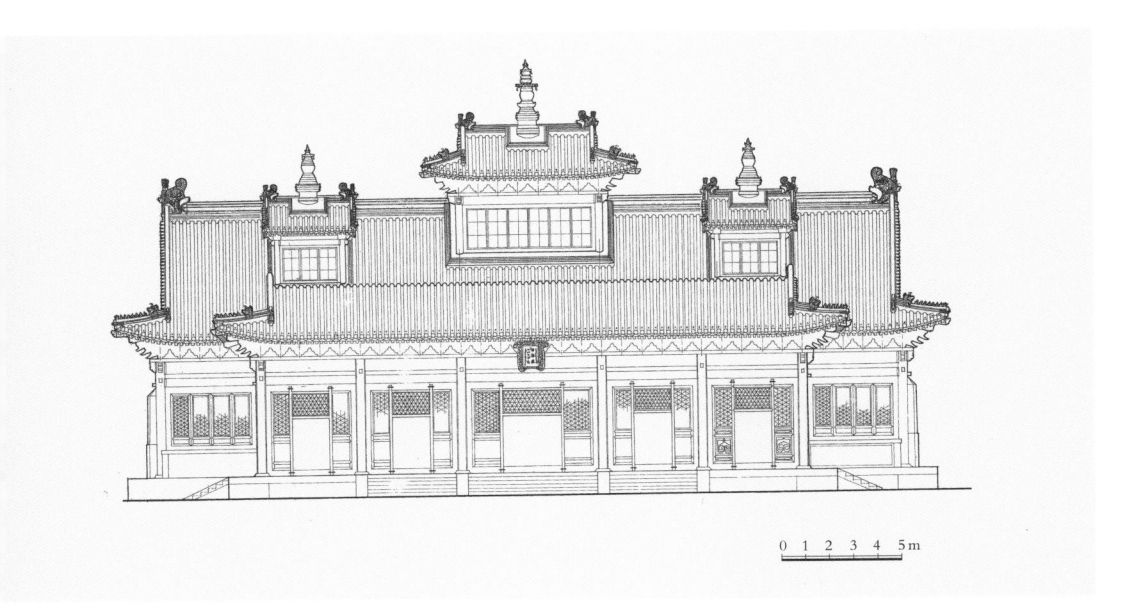

▲ Elevation of the Falun Hall

▲ Five Hundred Arhats in
the Falun Hall

▲ Interior View of the Falun Hall

Buddhist Temples

167

The fifth and also the last hall, the Wanfu Pavilion (Pavilion of Ten Thousand Happiness), also called the Dafolou Hall (Big Buddha Hall), was located at the site of the Rearmost Hall during the period when the Yonghegong Temple served as the emperor's residence. As the tallest structure in the Yonghegong Temple, the seven-bay pavilion is over 30 meters tall with a triple-eaved gable-and-hip roof. Seen from outside it is a three-story building. But seen from inside it is a tall pavilion without being separated by floor slabs. Inside the hall erects a 26 meters tall and eight meters wide Buddha statue carved from a single piece of sandalwood. This statue is one of three unique artworks in the temple. On the two sides of the Wanfu Pavilion are the Yongkang Pavilion and the Yansui Pavilion. The three pavilions are connected by double passageways up high in the air, looking like the Asgard. The pavilion contains the Buddhist architectural features of the Liao and Jin Dynasties and is the only existent building example in China. The east side hall of the Wanfu Pavilion is the Zhaofo Pavilion where Emperor Qianlong's mother used to worship. A bronze-glazed Buddha statue called "Sandalwood Buddha" was enshrined in the hall. Behind the Buddha stands a folding screen. The screen and the niche for the statue of the Buddha were carved with Phoebe nanmu. They were exquisite and unsurpassed and were considered as one of the three unique treasures in the Yonghegong Temple. The structure of the west side hall is the same as that of the east side hall. Behind the Wanfu Pavilion is the Rearmost Building called the Suicheng Building.

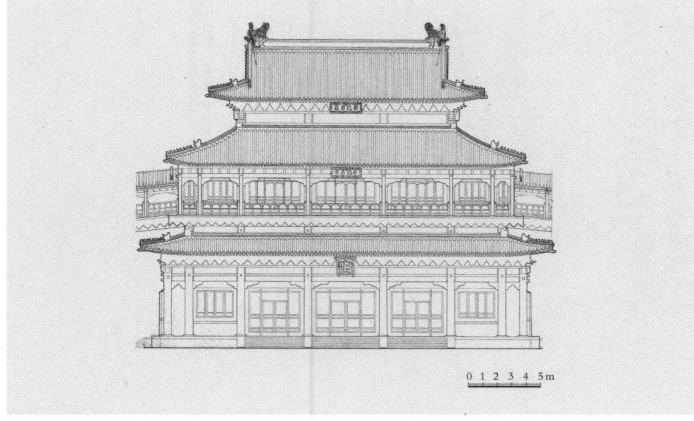

▲ Elevation of the Wanfu Pavilion

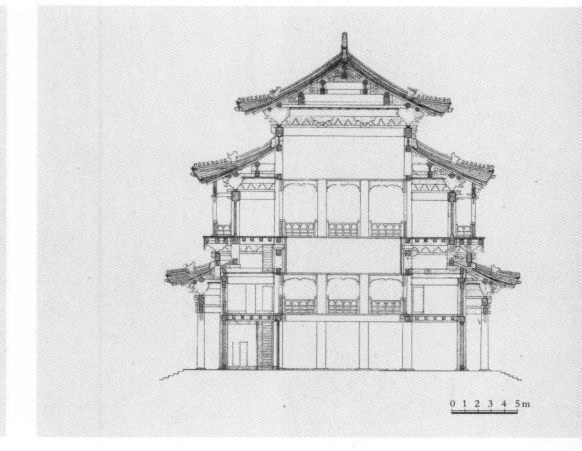

▲ Sectional Drawing of the Wanfu Pavilion

▲ The Wanfu Pavilion

▲ Sunk Panel of the Wanfu Pavilion

On the east axis stand the Academy of Classical Learning, the Safe and Sound Residence, the Good Luck Room, the Taihe Abstinence Hall, the Crabapple Courtyard and the garden etc. The structures on the west axis originally comprised the Goddess of Mercy Hall and the Temple of Guanyu. In addition, six residences for the Ajia Living Buddhas are located on the two sides of the imperial passageway in the front of the Zhaotai Gate. The living Buddhas in various dynasties all lived there.

Buddhist Temples

169

▲ Sandalwood Buddha in the Wanfu Pavilion

▲ Double Passages Connecting the Wanfu Pavilion to Other Pavilions

▲ Buddha Statues in the Zhaofo Pavilion

▲ The Suicheng Building

Buddhist Temples

# The Songzhu Temple and the Zhizhu Temple

The Songzhu Temple and the Zhizhu Temple are located at the North Songzhuyuan Lane in Jingshan Back Street, Dongcheng District. There were once three terrace temples: the Fayuan Temple in the east, the Songzhu Temple in the middle and the Zhizhu Temple in the west. Most of the structures in the Fayuan Temple are not existent now while the main structures of the Songzhu Temple and the Zhizhu Temple are still well preserved. It was designated by Beijing Municipal Government as a major historical and cultural site to be protected in 1984.

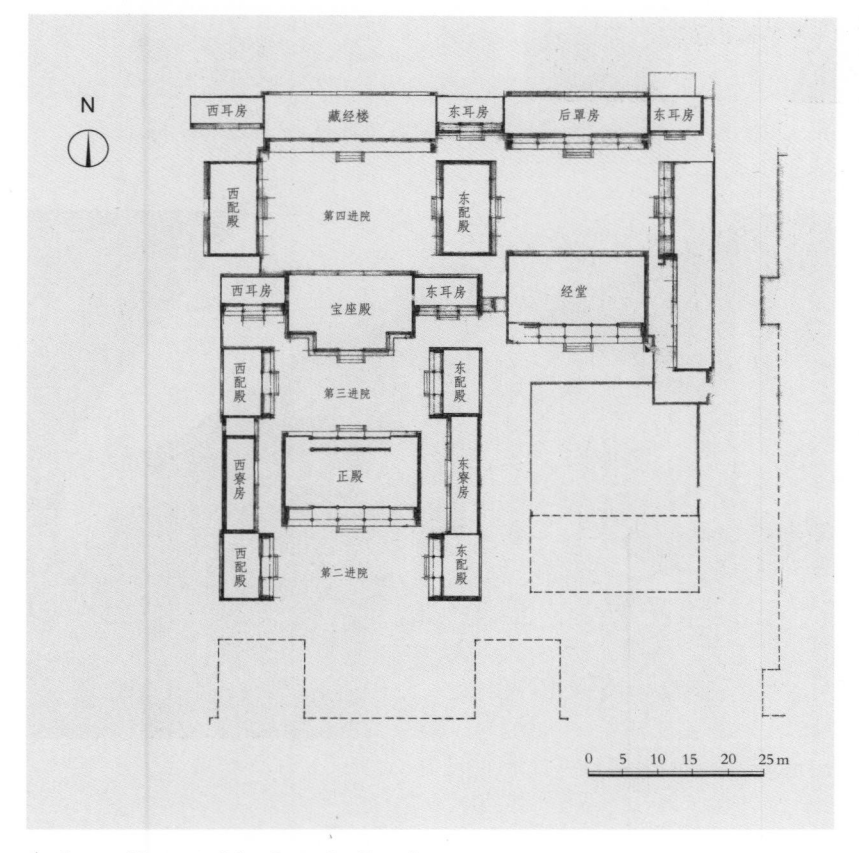

▲ General Layout of the Songzhu Temple

▲ The Main Hall of the Songzhu Temple

▲ Elevation of the Main Hall of the Songzhu Temple

Buddhist Temples

173

The south-facing Songzhu Temple are divided into three parts. The main structures are located on the central axis. There were originally five halls located on the central axis from the Main Gate to the Pavilion for Storing Buddhist Scriptures. The main gate, the Hall of the Heavenly Kings and the Bell and Drum Towers are non-existent now. The five-bay main hall has a veranda in its front and a flush gable roof supported by a main ridge and covered with round tiles. In the outer room of the main hall hangs a horizontal board on which were inscribed characters "Miao Ming Zong Jing". Three side halls are situated in the east and west sides of the main hall. The five-bay Throne Hall has a flush gable roof supported by a main ridge and covered with round tiles. In front of the hall stretch out three verandas with an overhanging gable roof supported by a hooped ridge and covered with round tiles. The portion under the eave was painted with tangent circle patterns. The Throne Hall was flanked with three wing halls in its east and west side. The two-story Pavilion for Storing Buddhist Scriptures is seven bays wide with a flush

▲ Scriptures-storing Pavilion of the Songzhu Temple

▲ The Throne Hall of the Songzhu Temple

Buddhist Temples

175

gable roof supported by a main ridge and covered with round tiles. In it there is a board and two couplets all written by emperor Qianlong. Only one set of courtyard is preserved on the east and west roads. On the east axis stand monks' dormitories, side rooms, Buddhist halls and the scripture halls, etc. The structures on the west axis mainly function as the Lamas' residences, among which only some of the buildings in the back are still extant.

The Zhizhu Temple was constructed from the 16th year to the 39th year (1751-1774) in Emperor Qianlong's reign. The south-facing temple has a three-bay main gate with a flush gable roof supported by a main ridge and covered with round tiles. There is an entrance door and a red enclosure in front of the main gate. On the stone board were inscribed the characters "Chi Jian Zhi Zhu Si" (Construction of

▲ Vaulted Doors in the Wall of the Scriptures-storing Pavilion of the Songzhu Temple

▲ The Main Hall of the Zhizhu Temple

the Zhizhu Temple by Imperial Command). The Bell and Drum Towers were torn down. The three-bay Hall of the Heavenly Kings has a flush gable roof supported by a main ridge and covered by round tiles. The roof was decorated with dragon-head ridge ornaments. Under the eave was decorated a Sancai single-ang bracket set that was painted with color tangent circle patterns. On the board in the hall were inscribed the characters "Bao Wang Guang Yin". Behind the Hall of the Heavenly Kings stands the square Main Hall. With a gallery in front of it, the Main Hall of three bays wide and deep has a double-eaved four-angled pavilion roof. The part right under the upper eave was decorated with a Wucai single-ang lifted bracket set. The lower eave column was decorated with a decorated bracket set and colored tangent circle patterns. The five-bay Jingshen Hall (Body Purification Hall) has a gable-and-hip roof supported by a main ridge and covered with round tiles. The part right under the eave was painted with tangent circle patterns and adorned with a Wucai single-lever bracket arch. On the lintel of the outer room hangs a board on which were inscribed the characters "Xian Qing Jing Shen", so the hall was named the Jingshen Hall (Body Purification Hall). With a veranda in front of it, the five-bay Back Hall has a flush gable roof supported by a main ridge and covered with round tiles.

Buddhist Temples

# The Main Hall of the Pudu Temple

Located at No. 35 Front Puqing Lane in Dongcheng District of Beijing, the Pudu Temple is a Tibetan Buddhist structure with a distinctive Manchu style. The temple was designated by Beijing Municipal Government as a major historical and cultural site to be protected in 1984.

The Pudu Temple was originally built on site where Dongyuan was constructed in the Ming Dynasty. It was in Dongyuan where the Ming Emperor Yingzong, Zhu Qizhen, was once detained. After he regained his reign he carried out a large-scaled renovation on the Dongyuan and named it Xiaonancheng (The Small South City). It was destroyed by the Peasants' War at the end of the Ming Dynasty. It served as Prince Ruiqin Dorgon's royal residence in the early years of the Qing Dynasty, which was called the Mansion of the Prince Regent. Emperor Kangxi ordered to transform the residence of Prince Ruiqin into the Maha Karma La Temple in the 33rd year (1694) of his reign. It was named "the Pudu Temple" by imperial command in the 40th year (1755) during Emperor Qianlong's reign and its main hall was named "the Ciji Temple". Beijing Municipal Government conducted an overall renovation on the temple which was finally completed in 2003.

In the front of the south-facing Pudu Temple stands the main gate which is

▲ The Main Hall

▲ Ornamental Front Eave of the Main Hall　　▲ Base of the Main Hall

three bays wide and seven purlins deep. The gate has a flush gable roof supported by a main ridge and covered by glazed green tiles. A white-stone archway was built under the front eave of the outer room while the side rooms were built with arched white-stone windows. The stone carvings on the windows were decorated with flower-pattern fans. A gallery stretched out in front of the back eave. The outer room was paneled with slab doors. The decoration of the side room is identical to that of the front eave. The interior of the main gate was painted with golden dragon patterns. A platform stands in front of the main gate and a brick-paved path behind the main gate which leads straightly to the main hall.

On the decorated stone base erects the main hall which is seven bays wide and nine purlins deep with a single-eaved gable-and-hip roof covered with round gray tiles and decorated with sheared and glazed green edges. The gallery, facade and the back of the hall were built with big removable windows whose sills were inlaid with six square glazed green tiles. Under the front eave of the outer and side rooms stand three verandas with a gable-and-hip roof covered with glazed green tiles and decorated with sheared yellow edges. The exterior eave in this hall has an overhang of three-tiered rafter and the top of the rafter was decorated with timber carvings of beasts' heads patterns. The shape of the decorated bracket is very peculiar. Some of the indoor color paintings were based on some ancient legends. The decoration of the whole structures shows distinct features of the Manchu royal palaces outside Shanhaiguan Pass. The rank of the tiles covered the cabinet is higher than that covered the main hall. This hall stores a revolving sutra cabinet discovered in 2002 when the temple was under renovation. On the west side of the main hall was preserved a north room of the abbot house. It is five bays wide and seven purlins deep with a gallery stretching out in front of the room that is topped by a flush gable roof supported by a raised ridge and covered with round tiles.

# The Fuyou Temple

Located at No. 20 Beichangjie Street in Xicheng District of Beijing, The Fuyou Temple is a very important temple in the imperial city. It was designated by Beijing Municipal Government as a major historical and cultural site to be protected in 1984.

First constructed in Emperor Shunzhi's reign in the Qing Dynasty, the Fuyou Temple was the place where young prince Xuanye (Emperor Kangxi) was taken to stay when smallpox was rampant. The emperor also studied there during his childhood. The main hall was constructed in the 1st year (1723) during Emperor Yongzheng's reign in the Qing Dynasty. The hall was given to Prince Baoqin (the title of Emperor Qianlong when he was still a prince), Hongli, as his residence but he didn't live in it. It was later transformed into a Lamasery and renamed the Fuyou Temple.

The south-facing Fuyou Temple has an outer westward wall door serving as the passage of the Temple. After entering the outer wall door, visitors will see a three-bay, four-column and seven-story Pailou. An 18 meters long one-glyph-

▲ The Pailou

▲ The Glazed Screen Wall

▲ The Main Gate

shaped screen wall stands at the south end of the temple. The three-bay main gate of the temple has a single-eaved gable-and-hip roof covered with glazed round yellow tiles. Imperial gateways carved with dragons were paved in front of and behind the main gate. Two splay screen walls were built on the left and right sides of the main gate. On the east and west sides behind the main gate were built the two-story Bell and Drum Towers with gable-and-hip roofs covered with glazed round yellow tiles with sheared green edges. The three-bay Hall of the Heavenly Kings has a single-

▲ The Hall of the Heavenly Kings and the Mahavira Hall

▲ Part of the Statue in the Hall of the Heavenly Kings

eaved gable-and-hip roof covered with glazed round yellow tiles with sheared green edges. On the east and west sides of the hall stand three side halls. Behind the Hall of the Heavenly Kings erects the five-bay Mahavira Hall which has a single-eaved gable-and-hip roof covered with glazed yellow tiles. The middle of the ridge ends was decorated with a Tibetan pagoda. In the front of the main hall stands a platform with an outstretched wall on its south, east and west sides. Behind the main hall was paved a path leading to the five-bay Rear Hall which has a single-eaved gable-and-hip roof covered with glazed round yellow tiles. The memorial tablet, the "Tablet to Memorize the Great Achievements of the Humane Emperor Sheng" was once enshrined in the hall. The Rear Hall has three side halls on its east and west and is flanked with the east and west side halls. The last structures in the temple are three rooms behind the Rear Hall where Buddhist scriptures are stored.

▲ The Mahavira Hall

Buddhist Temples

▲ Copper Tower Ridge Ornament
of the Mahavira Hall

▲ The Rear Hall

183

Divided into the Quanzhen School and the Zhengyi School, Taoism has a wide spread influence in Beijing. In the process of development, Taoism assimilated many Folk Gods worshiped by the folk people, placed many ordinary persons among Gods and had formed a huge Taoist God system. This chapter first introduces one of the three Zuting of the Quanzhen School — the White Cloud Temple — and the highest-leveled Taoist Temple — the Dagaoxuan Temple. Secondly it presents the Taoist temples represented by five temples including the Dongyue Temple that reflects the Dongyue God System. Furthermore, it also introduces four Taoist temples that are used to enshrine natural Gods and are represented by the Fire God Taoist Temple.

Taoist Temples

# The White Cloud Temple

The White Cloud Temple is located outside the Xibianmen in Xicheng District of Beijing. It is one of Beijing's oldest and largest Taoist temples, one of "The Three Great Ancestral Courts" of the Quanzhen School of Taoism, and is titled "The First Temple under Heaven." With valuable cultural relics, it is a great treasure-house of Taoism. It was designated by the State Council as a major historical and cultural site under state protection in 2001.

Originally called the Tianchang Temple (the Temple of Celestial Perpetuity), the White Cloud Temple has a history of over 1,200 years. Emperor Xuanzong in Tang Dynasty, named Li Hongji ordered to build a temple named Tianchang Temple in the 10th year under his reign (722) and bestowed a stone statue of Lao Zi to the temple. In Jin Dynasty, the temple experienced a war and emperor of Jin Dynasty ordered to rebuild the temple and renamed it the Shifang Da Tianchang Temple in the 7th year under his reign (1167). When destroyed by a fire in the 3rd year under the reign of Emperor Taihe of Jin (1203), the temple was reconstructed again and renamed the Taiji Palace (Great Ultimate Palace). In 1224, when Qiu Chuji returned from the East after visiting Chinggis Khan, he came to live in the Taiji Palace under the emperor's order and renamed it Changchun Palace (Palace of Eternal Spring). When Qiu Chuji passed away, his successor built a memorial shrine named Chushun Hall over Qiu's grave in the east of Changchun Palace. This shrine is today's Qiuzu Palace. In late Yuan and early Ming Dynasties, Changchun Palace was destroyed by the fire again. During the years under the reign of Emperor Yongle, the palace was moved to the east, was expanded with the Chushun Hall as its center and was renamed the White Cloud Temple. In the 1st year under the reign of Emperor Kangxi (1662), the Quanzhen master Wang Changyue made a large-scale renovation to the White Cloud Temple upon the imperial command. The renovation ended in the 45th year (1706) and laid a foundation for today's halls in the central road. In the middle period under the reign of Emperor Guangxu in Qing Dynasty, the 20th abbot of the White Cloud Temple Gao Rentong (Dharma name: Yuntong; alternative name: Yunxi) knew about Xuan Record and was proficient in painting and poetry, and was specially favored and trusted by Empress Dowager

Cixi along with his Taoist friend Liu Chengyin (Dharma name: Suyun) who is the imperial eunuch. Cixi's favor and trust made the influence of Quanzhen School reach as far as the imperial palace. Quanzhen School became very powerful in that period. The White Cloud Temple was renovated in 1924 again. After the founding of the Republic of China, it was renovated by the government for several times. At that time, the Taoist Association of China, the Taoist Institute of China and the Taoist Culture Researching Institute were established within the temple. Besides, the herbal medicine made by the Taoist priests within the temple was very popular among the common people.

The south-facing White Cloud Temple is situated on a north-south axis, with the roads and gardens on the eastern and western sides.

The main hall is situated in the central road. In the forefront is a one-glyph-shaped brick screen wall, engraved with four Chinese characters—Wan Gu Chang Qing (evergreen and everlasting) written by Zhao Mengfu, a calligrapher in Yuan Dynasty. Behind the screen wall is a four-pillar, three-room and seven-story Pailou (the decorated archway) built in the 8th year under the reign of Emperor Zhengtong in Ming Dynasty (1443). It was initially used for the Taoist priests to observe the

▲ The Pailou

▲ The Screen Wall

▲ Inscribed Board above the Main Gate

▲ Stone Monkey on the Screen Wall of the Main Gate

sun, the stars and the clouds. Behind the Pailou, the temple gate is seen. It is the main gate of the Taoist temple and was built in the 8th year under the reign of Emperor Zhengtong in Ming (1443). It separates the temple from the mortal world. The temple gate is of three bays wide with a beamless palace, gable and hip single-eave roof and covered with round tiles. Under the eave hangs a cast iron board engraved with "Chi Jian Bai Yun Guan" (Construction of the White Cloud Temple by Imperial Command). On both sides of the temple gate are stone lions and the white marble ornamental columns showing that temples were built upon the imperial orders and cannot be seen in ordinary temples. There is an allusion about the temple gate of White Cloud Temple, i.e. touching the stone monkeys. There are three stone monkeys located in different places inside the temple. It is said that if one can find all of them and touch them, the sufferings and disasters of the whole year can be eliminated. Behind the temple gate is a stone bridge called the Wofeng Bridge, under which there is no water. Going through the bridge, one comes to the first hall named Lingguan Hall of three bays wide, which is also called Four General Hall because the four generals Ma Lu, Zhao Gongming, Wen Qiong and

▲ The Main Gate

Guan Yunchang are enshrined in the hall during the years under the reign of Emperor Zhengtong in Ming Dynasty. When it was renovated in the years under the reign of Emperor Kangxi in Qing, it began to mainly sacrifice to Wang Lingguan who is the safeguard god of Taoism and was renamed Lingguan Hall. On the western and eastern side of the hall are two wing halls called Lingshui Hall and Shifang Hall. The bell tower and drum tower are on the north of Lingguan Hall with the former in the west and the latter in the east. The locations of the two towers are different from the Buddhist temple's layout of "bell tower in the east and drum tower in the west".

Behind the Lingguan Hall is the

▲ The Drum Tower

▲ The Wofeng Bridge and the Lingguan Hall

Yuhuang Hall, which was initially named the Yuli Changchun Hall. It is of five bays wide, and has flush gabled single-eave roof and covered with round tiles, platform are in front of the central bays and sub bays. The Jade Emperor is enshrined in the hall. The east wing hall is the Three Officials Hall, originally named "the Fengzhen Hall", in which are enshrined the three god officials of Heaven God, Ground God and Water God; the west wing hall is the Fortune God Hall, originally named "the Ruxian Hall" where three gods of fortune are enshrined. Behind the Yuhuang Hall is the Laolv Hall, initially named the "Qizhen Hall". The patriarch Wang Changyue in Qing Dynasty once conveyed the Taoist doctrines and transmitted precepts here upon the imperial order. The people all over the country wanted to learn the precepts and Taoism became very popular. Later generations renamed it "Laolv Hall" to memorize this prosperous period. The Laolv Hall is the place for the Taoism priests to perform religious activities. It is of a very large size, three bays wide and 15 purlin deep. It is of the connecting form with gable and hip single-eave roof and covered with round tiles. Within the hall, the seven successors of the patriarch of Quanzhen School Wang Chongyang i.e. "Qizhen in the North" are enshrined. The

▲ Interior View of the Yuhuang Hall

▲ The Yuhuang Hall

east wing hall of the Laolv Hall is Jiuku Hall (Suffer-relieving Hall) where Taiyi Jiuku Buddha is enshrined; the west wing hall is the Yaowang Hall (the Medicine King Hall) where the Chinese ancient famous doctor Sun Simiao is enshrined. The Qiuzu Hall is behind the Laolv Hall. It is the center of the buildings in the White Cloud Temple, originally the sub-house of Changchun Palace. Initially named Chushun Hall, it was built in 1228. In Ming Dynasty, the successors of Quanzhen School began to worship the patriarch Qiu Chuji here and named it the Changchun Palace; in early Qing Dynasty, it was renamed the Zhenji Hall; in Qianlong years of Qing, it was renamed the Qiuzu Hall and the name has been used ever since. The hall is three bays wide with flush gabled single-eave roof and covered with round tiles. The statues of the immortal Qiu and his two attendants are in the hall, and the scenes of the immortal Qiu and the first emperor of Yuan Genghis Khan meeting each other were carved in relief on the four walls. A Ying bowl which is said to be inherited from Song Dynasty is set on the goiter stone seat. It is bestowed by Emperor Qianlong to the White Cloud Temple. On the surface of the bowl is engraved "the 21st year of Emperor Qianlong's reign" and under the bowl is

BEIJING ANCIENT ARCHITECTURE SERIES

TEMPLES AND MOSQUES

▲ The Laolv Hall

▲ The Qiuzu Hall

▲ Ying Bowl in the Qiuzu Hall

194

the patriarch Qiu's body. Originally named the Sanqing Hall, the Sanqing-Siyu Pavilion is the last hall on the mid axis. It is five bays wide with a front gallery and enshrines statues of the Three Pure Ones. It was built in the 3rd year under the reign of Emperor Xuande in Ming Dynasty (1428) and rebuilt into a two-story structure where the Three Pure Ones are on the top floor and the Jade Emperor on the bottom floor. In the 51st year under the reign of Emperor Qianlong (1786), it was renamed Sanqing Siyu Pavilion with statues of the Three Pure Ones on the top floor and the Four Celestial Aides on the bottom floor. The pavilion is flanked by corner wing buildings, the Scriptures Storing Pavilion in the east, the Chaotian Tower (the Towards Heaven Hall) and the Moon Watching Tower in the west.

On the east road of the White Cloud Temple are the Gongde Hall (the Merits and Virtue Hall) for recording the donations, the Sanxing Hall (the Three Gods Hall) enshrining Gods of Fortune, Wealth and Longevity, the Cihang Hall enshrining the statue of Avalokitesvara Bodhisattva, the Zhenwu Hall enshrining the statue of Emperor Zhenwu, the Leizu Hall (the God of Thunder Hall) enshrining the statues of God of Thunder and four Buddhas, the Abstinence Hall for the Taoist priests having dinner. Besides, there is a three-level close-eave type octagonal pagoda with pyramidal roof, called the Luogong Pagoda where Luo's remains are preserved.

On the west road, there are totally six palace halls and three ancestral halls,

▲ Interior View of the Sanqing-Siyu Pavilion

▲ The Sanqing-Siyu Pavilion

with the seventh patriarch of Quanzhen Longmen School Wang Changyue enshrined and his remains were preserved under the hall. The walls inside the hall were inscribed with *Tao Te Ching* and *Yin Fu Ching* written by the well-known calligrapher Zhao Mengfu in the Yuan Dynasty. The Baxian Palace (the Eight Immortals Palace) is three bays wide with the eight Taoist immortals enshrined—Zhong Liquan, Lv Dongbin, Zhangguolao, Cao Guojiu, He Xiangu, Lan Caihe, Tie Guaili and Han Xiangzi. The Lvzu Palace is three bays wide with the Chunyang patriarch Lv Dongbin enshrined. It is the only palace of the green glazed tile roof in the White Cloud Temple. Initially named "Zisun Hall", the Yuanjun Palace is three bays wide. It is also known as Palace of Goddess, with the Goddess Bixia Yuanjun enshrined in the middle seat, the Sight Goddess and Smallbox Goddess on the right seats and Children-senting Goddess and Children-delivering Goddess on the left seats. Initially named "North Wuzu Palace", the Wenchang Palace is three bays wide, with Wenchang God who is in charge of fortune and fame enshrined in the middle, Confucius and Zhu Xi enshrined on both sides. The Yuanchen Palace is five bays wide and also known as "Sixty-year Circle Palace." It was built under the reign of Emperor Zhangzong in Jin Dynasty as a present for his mother with the Empress'

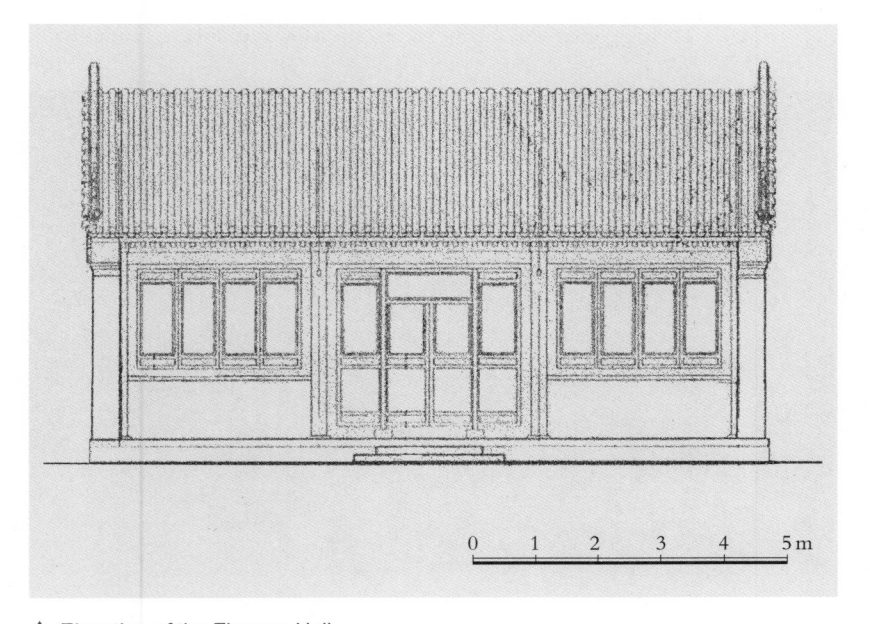

▲ Elevation of the Zhenwu Hall

▲ The Thunder God Hall

birth god enshrined and was named "Dingmao Ruisheng Palace" then. Now Doulao Yuanjun is enshrined in the middle of the palace, surrounded by the sixty-year circle gods.

The northern end of the White Cloud Temple is the back garden, named "Yunji Park", also known as "Xiao Penglai" (a small fabled abode of immortals). It was built in the 16th year under the reign of Emperor Guangxu (1890). The garden which is composed of three courtyards is quiet and beautiful and has many winding corridors, quite different from the solemn and splendid palaces in the

▲ The Luogong Pagoda

front of the temple. Just as the abode of fairies and immortals that is described in the Taoist doctrines, there is a unique atmosphere in the back garden. Inside the garden, there are Youhe Pavilion and Yunhua Celestial Hall in the east, Miaoxiang Pavilion and Retreat Hall in the west and the precept platform and Yunji Mountain House in the middle. The north-facing Yunji Mountain House is the central building of the garden, five bays wide with galleries all around. It is the place for the Quanzhen School patriarch to give precepts to the believers. The precept platform is the place for the Quanzhen School patriarch to give "the three precepts" including Chuzhen precept, Zhongji precept and Tianxian precept. In 1989, the White Cloud Temple held the first precept-giving ceremony by Quanzhen sect ever since the founding of the Republic of China. There is a rockwork in each courtyard, signifying the three immortal mountains in Taoism, i.e. Penglai, Yingzhou and Fangzhang. The whole garden is composed of the various procedures to become immortal: getting out of the mortal world, retreating to cultivate oneself according to the Taoist doctrines, receiving precepts in the precept platform, thinking about the Taoist doctrines, entering the three immortal mountains' abode and residing in Yunhua Celestial Hall, so it can be regarded as a typical example of the temple garden.

▲ The Lvzu Hall

▲ Te (A Legendary Beast, like a Donkey and a Horse)

▲ The Retreat Hall

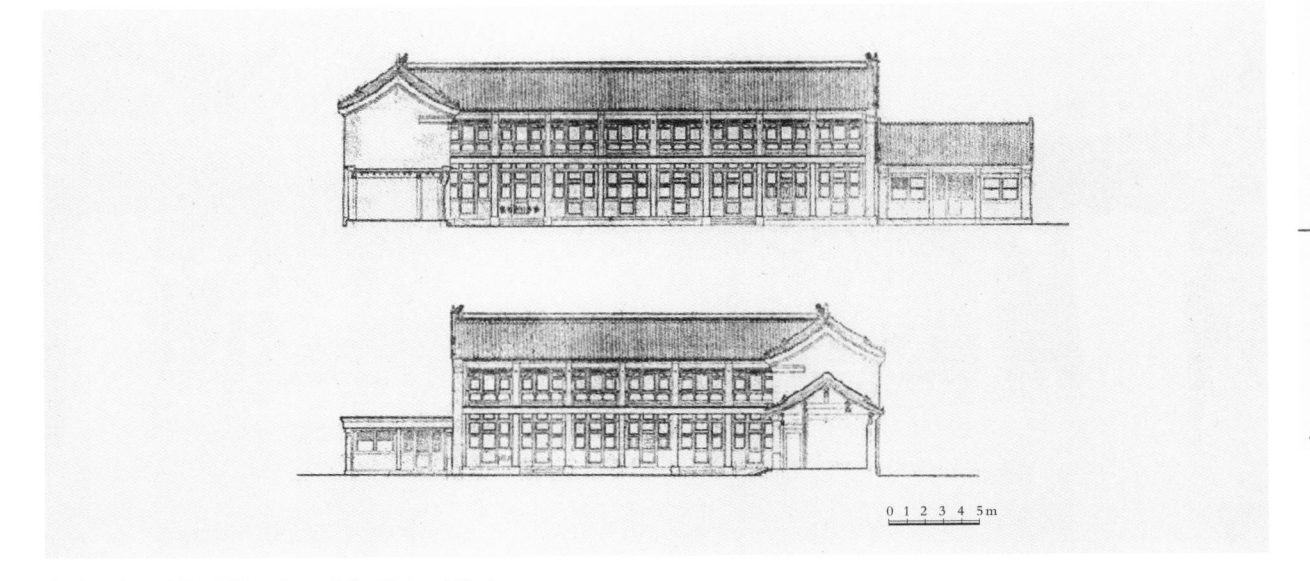

▲ South and East Elevation of the Retreat Hall

Besides, many upright stone tablets erected since Ming and Qing Dynasties are preserved in the White Cloud Temple. Some of the important ones are: the tablet of "Renovation of the White Cloud Temple" written by Hu Ying in the 9th year under the reign of Emperor Zhengtong in Ming(1444), the "Tablet of Bestowed Scriptures" written by Xu Bin in the 13th year under the reign of Emperor Zhengtong in Ming Dynasty (1448), the "Tablet of Taoist Virtue of Changchun Immortal Qiu" written by Zhang Zan in the 1st year under the reign of Emperor Zhengde (1506), the "Tablet of Renovation of the White Cloud Temple" in the 13th year under the reign of Emperor Zhengde (1518), the "Tablet of Renovation of the White Cloud Temple" written by Gu Yishou in Emperor Jiajing' year and "The Record of the Renovation of the White Cloud Temple" written by Wang Changyue during the years in Emperor Kangxi's reign in Qing Dynasty, the "Record of Renovation upon the Imperial Command" and the "Poetry Tablet Erected upon the Imperial Command" in the 53rd year under the reign of Emperor Qianlong in Qing Dynasty(1788), etc.

▲ The Miaoxiang Pavilion

▲ The Yunji Mountain House

▲ The Precept Platform         ▲ The Gallery and the Inscribed Stele        ▲ The Yuxian Pavilion

Taoist Temples

▲ The Yunhua Celestial Hall

201

# The Dagaoxuan Temple

The Dagaoxuan Temple is located at No.21 and No.23 Jingshan West Street, Xicheng District, Beijing. It was the only temple that was exclusive to the royal family in Ming and Qing Dynasties. With special architectural features, it was designated by the State Council as a major historical and cultural site under state protection in 1996.

The Dagaoxuan Temple which was the abstinence hall of Emperor Jiajing in Ming Dynasty was built in the 21st year under the reign of Emperor Jiajing (1542) and was destroyed in the fire in the 26th year under his reign (1547) and rebuilt in the 28th year under the reign of Emperor Wanli (1600). In order to avoid the taboo of Emperor Kangxi's imperial title, it was renamed "the Dagaoyuan Temple" and later renamed "Dagao Temple". It was renovated in the 8th year under the reign of Emperor Yongzheng in Qing Dynasty (1730), in the 11th year under the reign of Emperor Qianlong (1746) and in the 23rd year under the reign of Emperor Jiaqing (1818). In 1990 when the Eight-Power Allied Forces invaded China, the temple was seriously damaged and was repaired later. During the period of the Republic of China, some buildings in front of the temple were torn down or transformed. After the founding of the Republic of China, the temple was used by the army.

The south-facing Dagaoxuan Temple takes the shape of a rectangle. It covers an area of 13,000 square meters. In the forefront, there are three decorated

▲ Old Pictures of the Pailou and the Doctrines-learning Pavilion

archways, two doctrine-learning pavilions. The latter has a special structure of five-flower pavilion type with tripling eaves, gable and hip crossing ridge, just like the corner tower of the Forbidden City, but more delicate. These buildings were all torn down after 1949 because of the expansion of the streets. The southern decorated archway was rebuilt in 2004, taking the style of four-pillar, three-bay and nine-story. It was reconstructed in the way it was repaired in 1927. It is 10.08 meters high and 16.6 meters wide, wearing the yellow glazed roofing with a bracket set,

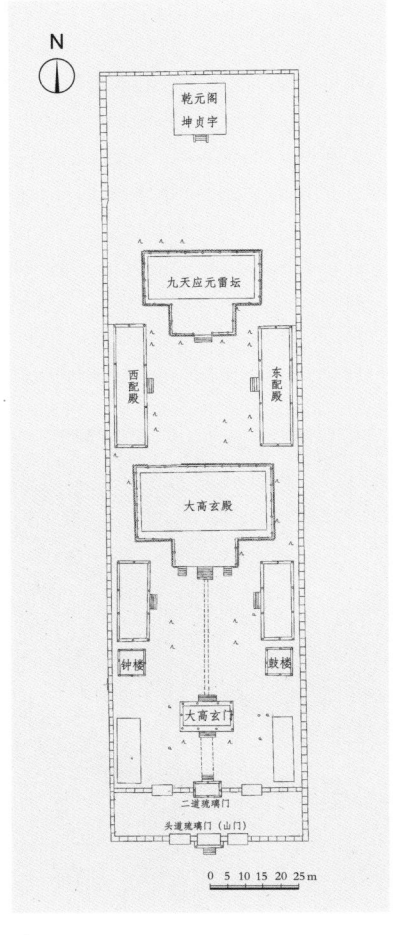

▲ General Layout of the Dagaoxuan Temple

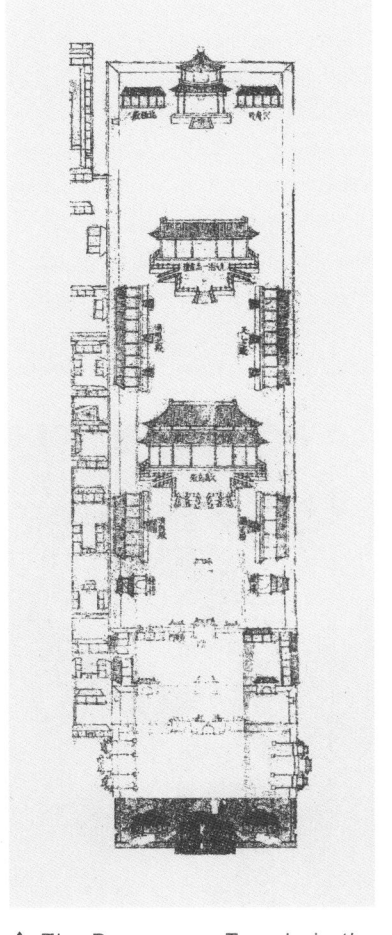

▲ The Dagaoxuan Temple in the *Panoramic View of Beijing under Emperor Qianlong's Reign*

tangent circle pattern of ink line under the eave. The northern inscription board reads "Da De Yue Sheng"( The most valuable virtue is to take care of all creatures), the southern one reads "Qian Yuan Zi Shi" ( Strength of all creatures comes from Qianyuan) and the bottom clamping lever stone was engraved "Shou Yu Tian Qi"(Longevity Forever). The shape of the west and east decorated archway is the same as the southern one but with different inscriptions on the board: the eastern and western inscription board of the eastern archway reads "Kong Sui Huang Zuo"(Kong Pacifies the Imperial Throne) and "Xian Tian Ming Jing" (The Bright Mirror); the eastern and western inscription boards of the western archway read "Hong You Tian Min"(Protecting the Common People) and "Tai Ji Xian Lin" (Taiji Gods' Abode). The temple gate of the Dagaoxuan Temple is opposite to the southern decorated archway, taking the form of three arch gates. It has gable and hip roof and green glazed tile roofing and is of glazed tile wood-simulating structure. The front gate has an inscription board on which were written "Shi Qing Dao Jing" (Starting the Quiet Taoist Abode). It is the only temple gate in Beijing that is of such a form. Going through the temple gate there is a glazed tile gate in the form of three gates, with the middle bigger and the two flanks smaller. It has gable and hip single-eave roof, green glazed tile roofing with a bracket standing under the eave. Behind the two gates is a Dagaoxuan gate in the form of the gallery of three bays. It

▲ The Main Gate

▲ The Dagaoxuan Gate

has a gable and hip single-eave roof. Five arches are under the eave. On both sides behind the Dagaoxuan Gate are a bell tower and a drum tower which have gable and hip double-eave roof and yellow glazed tile roofing with a bracket set of three liters under the eave. The bell and drum towers were also destroyed when the Eight-Power Allied Forces invaded Beijing. There was a flagpole in front of the gate and now only a stone seat is left.

Behind the Dagaoxuan Gate is the main hall called the Dagaoxuan Hall. The hall is situated on a Sumeru seat surrounded by white marble stone rails. In the front of the hall is a platform with three steps in the front. In the middle is a stone royal road engraved with the patterns of cloud-dragon, cloud-phoenix and crane. The hall is seven bays wide with double-eave hip roof and yellow glazed tile roofing. Above the eaves stands a bracket set of one flower arm petal, double levers and seven horizontal bracket arms and under the eaves is a bracket set of double levers

205

▲ The Dagaoxuan Temple

and five horizontal bracket arms, golden dragon patterns. The central bay and sub bay have four doors with four partition boards and the end bay has windows with caltrop flower patterns of three cross and six bows with bronze recto. The east and west wing halls are of five bays with front gallery, gable and hip single-eave roof and green glazed imbrex roofing. Under the eaves stand a bracket set of two liters in the shape of nettle leave and the tangent circle pattern. There isn't any temple in Beijing with the main hall and wing hall of such high-level structures.

Then it comes to the Jiutian Yingyuan Thunder Altar. Emperor Zhenwu was originally enshrined in the hall. In the front of the hall is a platform surrounded by white marble rails and in the middle is a royal road. The altar has single eave hip roof, green glazed tile roofing with yellow sheared edge. Under the eave stands

▲ Imperial Road of the Dagaoxuan Temple

▲ Sunk Panel of the Dagaoxuan Temple

a bracket of double levers and five horizontal bracket arms and the tangent circle patterns. The east and west wing halls have nine bays each with gable and hip single-eave roof of large ridge and green glazed tile roofing. The east wing hall is the Tianyi Hall and the west is the Tongming Hall.

The last building is a two-story pavilion named the Qianyuan Pavilion representing the cosmic view of "Heaven is round and Earth is square" in ancient China. It is a place for the emperors in Qing Dynasty to pray for the rain. The Jade Emperor is enshrined in the pavilion. The top of the pavilion is round; the bottom square. It is situated on a platform seat surrounded by white marble stone rails and with steps and stone royal road in the middle. The circular pavilion on the top is situated on a flat seat surrounded by corridors with wooden rails. The top of the circular pavilion is a round pyramidal roof composed of eight columns and is covered with blue glazed tiles, representing the heaven. The square Kunzhen Pavilion in the bottom is covered with yellow glazed tiles, representing the Earth. Under the eaves stand a bracket set of one flower arm petal and one lever. The patterns of golden dragons, heavenly flower and the sunk panel are very beautiful. The whole pavilion has a characteristic structure which is the only one among

▲ The Qianyuan Pavilion

Beijing's Taoist temples.

Among the Taoist temples in Beijing, the Dagaoxuan Temple is the temple of highest level in Ming and Qing Dynasties. It possesses unique features. It has a double eaved hip roof which is of the highest level and the distinctive five-flower ridge. Representing that "Heaven is round and Earth is square" the Qianyuan Pavilion is beyond comparison. All of these contribute to its uniqueness among all the temples in Beijing.

▲ Interior View of the Qianyuan Pavilion

Taoist Temples

# The Beijing Dongyue Temple

Located in the north of Chaoyangmenwai Street, Chaoyang District, Beijing, the Dongyue Temple is where Dongyue gods are enshrined. It is the largest temple of Taoist Zhengyi School in North China, containing a large number of Taoist statues and tablets in different dynasties. The Dongyue temple is Beijing's unique site. It is quite valuable in the research of the history of Chinese ancient Taoism and was designated by the State Council as a major historical and cultural site under state protection in 1996.

In the period under the reign of Emperor Yanyou, after Zhang Liusun who was the 38th generation of the founder of the Taoist Xuan School Zhang Daoling was conferred the patriarch of Xuan School by Emperor Yuan Chengzong, he realized that Dongyue Temple enshrining the emperor Dongyue were built everywhere all over the country except Dadu, the capital of Yuan Dynasty. So he purchased the land outside Qihuamen at that time and prepared to start the project. But he passed away shortly afterwards. His disciple, the Taoist master Wu Quanjie continued the construction, and the temple was completed in the 3rd year under the reign of Emperor Zhizhi of Yuan Dynasty (1323) and was named the "Dongyue Rensheng Palace" by the imperial court. In the 2nd year under the reign of Emperor Taiding (1325), the Princess from Lu supported the construction of the emperors' resting palace which was finished in the 1st year under the reign of Emperor Tianli (1328) and was named the Zhaode Palace. In late Yuan Dynasty, the temple was seriously damaged in the war.

In Ming Dynasty, Xuan School was incorporated into Zhengyi School of Taoism and the Dongyue Rensheng Palace was renamed the Dongyue Temple. In the 12th year under the reign of Emperor Zhengtong (1447) in Ming Dynasty, the temple was repaired at its original site. Then, during the reign of Emperor Jiajing and Longqing, the temple was renovated several times. In the 3rd year under the reign of Emperor Wanli in Ming, upon the order of the Empress Dowager, the temple was expanded with the support of government. In the 37th year under the reign of Emperor Kangxi in Qing Dynasty (1698), the temple was destroyed by fire and most buildings were demolished. In the 39th year under the reign of Emperor

Kangxi (1702), the temple was rebuilt upon Emperor Kangxi's order and it was completed in the 41st year under the reign of Emperor Kangxi (1702). In 1761 during the reign of Emperor Qianlong, the temple was repaired again. During the period under the reign of Emperor Daoguang, Ma Yilin, the abbot of the temple collected alms everywhere and built the two side yards in the east and west as well as over 100 houses. He also founded a school, accepting the poor students free of charge. The buildings extant are all Qing Dynasty architecture. The Dongyue Temple began to decline since 1900, and it could pay its daily expenditure only by renting its houses. After the founding of the People's Republic of China, the Dongyue Temple was occupied by the government. In 1995, Beijing Municipal Government decided to withdraw all the offices from the Dongyue Temple. Built in the temple, Beijing Folk Customs Museum was formally opened to the public in 1999.

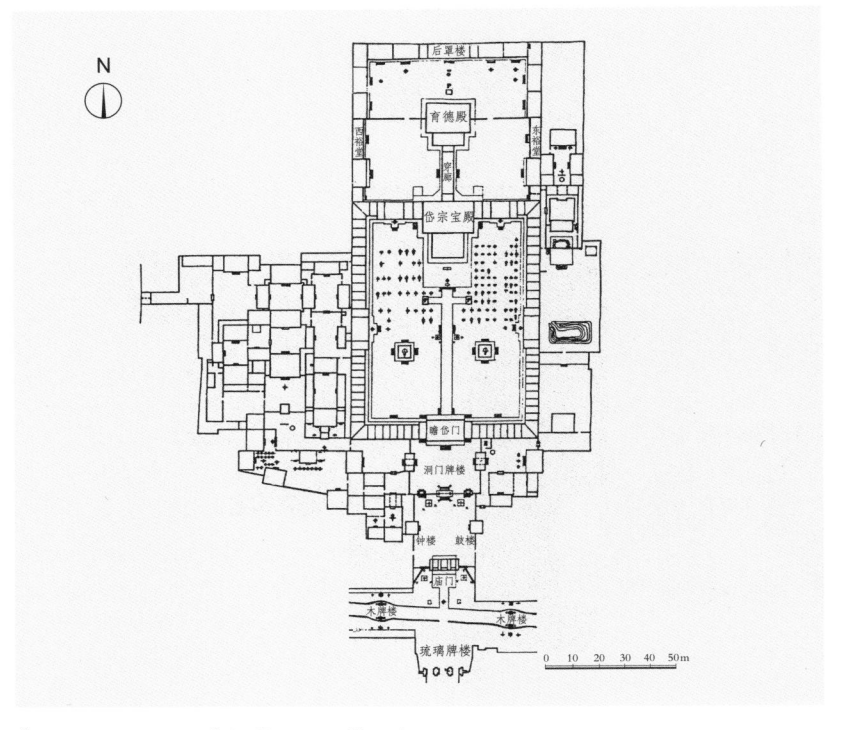

▲ General Layout of the Dongyue Temple

▲ The Glazed Pailou

▲ The Dongmen Pailou

The south-facing Dongyue Temple is composed of three parts: the central road and the eastern and western side yards. The grand main buildings of the temple are located in the central road. The main halls distributed on the south-north axis are (from the south to the north) the glazed arch, the temple gate (already removed), the Dongmen Pailou, the Zhandai Gate, the Daizong Hall, the Yude Hall and the Rearmost Hall, which are all flanked by secondary buildings. The fact that the main buildings are distinguished from the secondary ones, the front part echoes with the rear part and the symmetrical layout make the temple a true royal

Taoist temple.

The glazed tile arch was built in the 30th year under the reign of Emperor Wanli in Ming Dynasty (1602). It is a three-bay wide and seven-story structure with four pillars, gable and hip roof and green glazed imbrex roofing. The middle of the ridge is decorated with flaming bead; the eave is decorated with wood-simulating glazed brick bracket sets; above the architrave are the patterns of yellow and green composite flowers. On both sides are the white stone pillars with the central pillar of brickwork and the pillar root set with post stone. Between each two pillars are three arch doorways of bricks. Under the main building eaves are embedded two stone boards with the southern one reading "Zhi Si Dai Zong" (Worshipping Emperor Daizong for a Decade) and the northern one reading "Yong Yan Di Zuo"(The Throne Lasting Forever)which are said to be written by Chancellor Yan Song in Ming Dynasty. A wooden decorated archway was initially set up on both sides of the arch. They were torn down later.

The temple gate is the main gate of Dongyue Temple. It was pulled down in 1988 for expanding Chaowai Street. The temple gate was three bays wide with a gable and hip roof, grey imbrex roofing with green sheared edge and three bays' archways. A stone inscription board was embedded in the front door, reading

▲ The Zhandai Gate

213

"Dongyue Temple Built upon the Imperial Command".

Commonly referred to as the archway gate, the Dongmen Pailou is the second gate of the Dongyue Temple. After the temple gate was pulled down, it became the main gate. It has the hip roof and gray imbrex roofing with green sheared edge. The front door is one bay wide with a double gate and two supporting pillars on both the front and rear eaves. Since the Taoist temple is called the adobe of fairies and immortals by the Taoist believers, the gate is called "Dongmen Pailou" (the gate of the adobe). The inscription board of the "Dongyue Temple" written by Emperor Kangxi that was originally embedded on the temple gate was moved to Dongmen Pailou too. Inside the gate is the bell tower on the left and drum tower on the right. It has gable and hip roof and gray imbrex roofing with green sheared edge. Under the eaves is a bracket set. The top inside the gate is decorated with sun-block planks of arch windows and the bottom decorated with the eave cover of imbrex of green sheared edge. The four sides are all of brickwork and in the facade stands an archway. The bell tower in the west is inscribed with"Jing Yin"( The Sound of the Whale) and the drum tower in the east is inscribed with "Gui Yin"(The Sound of the Turtle) .

To the opposite of the Dongmen Pailou is Zhandai Gate which is five bays wide with a hip roof of single eave and gray imbrex roofing with green sheared edge. The outer bays and sub bays are the hallways, at the end of which the Un &Ah great generals and the ten Taibao are enshrined. Since the Un &Ah great generals are called immortal dragon and tiger general respectively, Zhandai Gate is also known as the Dragon and Tiger Gate. The beam frame and the structure inside the Zhandai Gate show the typical features of architecture of Ming Dynasty. Inside the Zhandai Gate, there is a pagoda tree in the west that has strong trunk and thin branches. It was said that the pagoda tree has a history of over 800 years, which owned it the name "Pagoda Tree of Longevity". Many people pray in front of this tree for longevity.

Behind the Zhandai Gate, there is a royal road called "Road of Fortune" which leads to Daizong Hall. The road of fortune is flanked by two pavilions with yellow grazed tile roofing, where the tablets inscribed by Emperor Kangxi and Qianlong are embedded. On the eastern and western side of the courtyard are two groups of galleries with 72 bays each, representing the 72 underground law officials in the charge of Emperor Dongyue. The buildings are of connected eave-straight ridge type with gray imbrex roofing. The bracket sets under the front and rear eaves all take the shape from Yuan Dynasty. On each lintel hangs the inscription board of the enshrined god; on the flank pillars pastes the couplet. The statues inside the hall are vivid both in form and spirit. Four officials were added to the original 72 statues of underground law officials, which make the total number reach 76. But all of the statues were removed later and the clay sculptures made by the successors of "Clay Figure Zhang" took their places in 1995 when the gate was rebuilt.

▲ Stele Pavilion

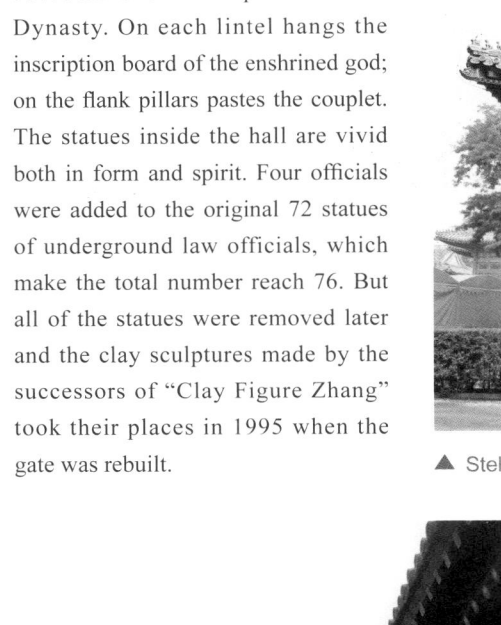

▲ Gallery of 72 Statues of Underground Law Officials

Taoist Temples

215

▲ The Daizong Hall

Daizong Hall is the main hall of Dongyue Temple. It is five bays wide with hip roof of single-eave and gray imbrex roofing. The corridor in front of the hall is three bays long with gable and hip round-ridge roof of end-painted ridge, gray imbrex roofing with green sheared edge. The beam, pillar, purlin and tie-beam in the hall are all painted with the golden dragon patterns which are exclusive to the royal family. Under the eave in the front hangs an inscription board, reading "Daiyue Hall" which is decorated with engraved dragons and golden leaves all around. Behind the Daizong Hall is a four-purlin veranda

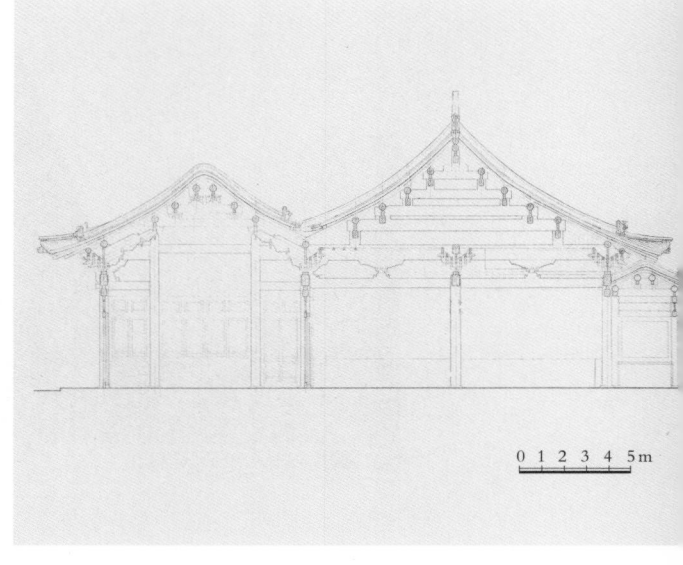

0 1 2 3 4 5m

▲ Sectional Drawing of the Daizong Temple

▲ Interior View of the Daizong Hall

▲ "Little Golden Beans" Stele

with overhanging gable roof of one bay which is connected with the corridor leading to the resting palace. In front of the hall is a platform with the bronze censer and five stone offerings on it. The platform is also flanked by two silk-burning censers. The statues of Emperor Dongyue and his attendants enshrined in the hall are no longer there. The main hall is flanked by two wing halls of three bays with gable and hip roof. The east wing hall enshrines the statue Three Immortals named Mao, while the west wing hall enshrines the statue of Binglinggong who is the god in charge of the three mountains. The east side hall is Fucai Hall of three bays and the west side hall is Guangsi Hall of three bays, both of which have gable and hip roof, gray imbrex roofing with green sheared edge. Under the eave of the front galleries of the side halls have bracket sets with connecting wood and tangent circle pattern of ink line, which show the Yuan Dynasty's style.

Behind Daizong Hall is the resting palace, they are joined by a corridor, under whose eave is a bracket set of two liters with the head in shape of nettle leave and the tangent circle pattern of ink line, which is the typical Yuan Dynasty's layout that is seldom seen in Beijing. The resting palace is called Yude Hall which is five bays wide and has hip roof and gray imbrex roofing with green sheared edge. Under the eaves stands a bracket set with one flower arm petal, double levers and five horizontal bracket arms. The corridor in front of the hall echoing with Daizong Hall is decorated with dragon and phoenix heavenly flowers' pattern. Inside the hall hangs an inscription board of "Xuan Miao Zan Hua" (Splendid Cultivation) written by Taoist priest Lou Jinyuan in Qing Dynasty. The statues of Emperor Dongyue and Empress Shuming Kunde originally enshrined in the hall were substituted by

▲ Gallery between the Daizong Hall and the Yude Hall

the statues of three gods and nine officials (the statues of three gods were formerly enshrined in Daci Yanfu Palace). The resting palace is flanked by the west and east side halls of three bays with gable and hip roof which are called Four Sons Hall. Four Sons Hall enshrines the four sons of Emperor Dongyue, i.e. Master Jingjian, Marquis Xuanling, Marquis Youling and Marquis Huiling.

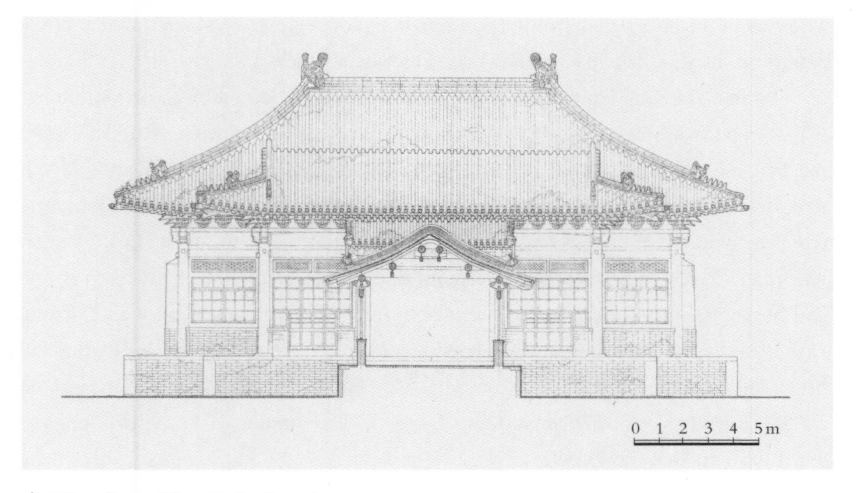

▲ Elevation of the Yude Temple

▲ The Yude Hall

The last building in the mid axis of the Dongyue Temple is the two-story Rearmost Hall which was originally called the Yuhuang Pavilion, the Bixia Yuanjun Hall, the Doumu Hall, the Da Xianye Hall, the Guandi Hall, the Zaojun Hall, the Wenchang Dijun Hall, the Xishen Hall, the Lingguan Hall and the Zhenwu Hall, etc. The Dongyue Temple was transformed into Beijing Folk Customs Museum. The structure has connected eave-straight ridge, gray imbrex roofing with green sheared edge and tangent circle pattern. In the west of the hall, there is a Yuzuo House for the emperors to rest when coming to the temple to worship or on the way to Dongling to worship the ancestors.

The east courtyard is the major residential area of the Dongyue Temple. The buildings here are scattered and have a living atmosphere. Inside the courtyard, there are corridors, wonderful trees and flowers, pavilions and fantastic rocks, which make it a beautiful garden. It was said that Emperor Guangxu and Empress Dowager Cixi often visited here.

The west courtyard is another worshipping area in the Dongyue Temple. It consists of various small yards enshrining all kinds of deities, including the Dongyue Hall, the Jade Emperor Hall, the Three Emperors Hall, the Medicine God Hall, the Xianhua Hall, the God of Horse Hall, the Miaofeng Mountain Goddess Hall, the Lu Ban Hall, the Three Gods Hall, the God of Plague Hall, the Yama Hall

▲ The Rearmost Hall

and Judges Hall, etc. They are of a relatively small scale and are mainly built with common people's sponsorship.

In addition to the buildings, the Dongyue Temple has three features that are noteworthy. Firstly, over 3,000 vivid god statues, among which there were the underground 72 law officials of different size and actions, are the masterwork in sculptures. So there was a saying that "the Dongyue Temple's god statue is the best all over the world."

Secondly, over 160 tablets were erected in each courtyard of the Dongyue Temple and they take the first place in Beijing. The over 100 tablets are of Yuan, Ming and Qing Dynasties. They are mainly the record of reconstruction of Dongyue Temple and the tablets from Common Confraternity. They are of various kinds, with noble sense, rich content and high artistic and historical value. The most well-known one is the running script of Zhao Mengfu—"The Taoist Tablet of Taoist Master Zhang" (commonly known as "The Taoism Tablet") which is the artistic treasure of Yuan Dynasty's calligraphy, unsophisticated and vigorous. Besides, the regular script of Zhao Shiyan—"Tablet of the Zhaode Hall", the clerical script of Yu Wenjing—"Tablet of the Rensheng Palace" are also very famous. but they are inexistent now. In addition, there are two very special tablets. One is from the 7th year under the reign of Emperor Shunzhi (1650). The tablet head is decorated with hollowed-out curled-up dragon pattern, so two persons who standing in the front and back of the tablet can see each other. The tablet is called "Transparent Tablet" or "Transparent Dragon Tablet". The other is in the east tablet group in the Dongyue Temple from the period under the reign of Emperor Shunzhi, named "The Record of Renovation of the Golden Lantern in the Dongyue Temple". On each side of the tablet seat is engraved a Taoist boy holding a lantern who are called "the naughty boys". Seen from different angles, they wear smile in their eyes. There was a tale about the pair of Taoist boys. The boys had been there for such a long time that they obtained a mortal spirit. One night they went out of the temple with the lanterns in their hands. The next day, a booth owner went to the temple and found the two boys were engraved on the tablet seat. Then he pasted the left boy's lantern with paper and tied one of the right boy's feet with a thread. From then on, no one has ever seen them go out of the temple.

Thirdly, couplets with black characters of small seal type on white background can be seen on both sides of the gates of most halls in the Dongyue Temple. Their content is usually about the explanation of the gods' responsibilities and persuasion which are thought provoking. The couplets in the front of the halls in the main courtyard were all renovated and written by the contemporary famous calligraphers.

▲ Stele Forest

# The Guangren Palace

The Guangren (Extensive Humanity) Palace is located in Lan Dianchang Village, Shuguang Street, Haidian District. It is one of Beijing's Wuding Temples. It was designated as a historical and cultural site to be protected by Beijing Municipal Government in 2001.

Wuding (five tops) refers to the five temples enshrining Bixia Yuanjun, the God of Mount Tai (the daughter of Emperor Dongyue). It was also named Bixia Yuanjun Temples, including the East Top, the West Top, the South Top, the North Top and the Mid Top. At present, only the West Top Guangren Temple is well-preserved. Second to it is the North Top which was renovated for the 2008 Beijing Olympics. All the other three temples were completely demolished.

The old site of Guangren Palace was built on the site of the Jiaxiang Taoist Temple built in Ming Dynasty under the reign of Emperor Zhengde. In the 10th year under the reign of Emperor Wanli (1582), the funds from the national treasury were

▲ The Main Hall

used to reconstruct the temple. The project was completed in the 18th year under the reign of Emperor Wanli (1590) and the temple was named by the Emperor as Huguo Hongci Palace (State-defending and Great Mercy Palace) which was renovated under the reign of Emperor Tianqi. In the 47th year under the reign of Emperor Kangxi in the Qing Dynasty (1708), the renovation was started and finished the next year. In the 51st year (1712), it was renamed the Guangren Palace. During the period of the Republic of China, it had been used as the Shezhou House (Porridge-Giving House). In 1949, it became the sanatorium and was transformed into Beijing Rubber and Hardware Factory in 1970s.

In the south-facing Guangren Palace are buildings including the three-bay palace gate of brick and stone with a gable and hip roof of gray round tiles; three arch gates connected with a splayed screen wall; the Pailou (decorated archway) in front of the palace gate which no longer exists; the three-bay Hall of the Heavenly Kings with the Four Heavenly Kings enshrined, under whom are the eight ghosts and monsters; the two wing halls in the east and west on both sides of the temple; the five-bay main hall, i.e. Goddess Hall, with a flush gable roof of green glazed round tiles and yellow sheared edge, under the front part of which there are five steps leading to the throne. Inside the hall is enshrined the statue of Bixia Yuanjun, on the left of which is the Yanguang Goddess (Eyesight Goddess), on the right of which is the Songzi Goddess (Children-sending Goddess) and under which are the warriors, maidservants and yaksha, etc. In front of the palace are several tablets with inscriptions of the Ming and Qing Dynasties, most of which describe pilgrimage to the temples. Besides, there are also the tablet of the "Inscriptional Record of Hongci Temple upon the Imperial Command" in the 4th year under the reign of Emperor Tianqi (1624), the re-erected tablet in the 47th year under the reign of Emperor Kangxi (1708) and the tablet built under the emperor's command in the 51st year under the reign of Emperor Kangxi (1721). The east and west wing halls are three-bay each with front aisles and are flanked by two wing halls on both sides. Inside the hall, there are the wall-paintings of the 74 officials from the underground world. In front of each official statue is an iron censer with the words "the 40th Year under the Reign of Emperor Wanli (1612)" on it. The five-bay Resting Hall wears flush gable roof of green glazed round tiles and yellow sheared edge. The main hall and the Resting Hall are connected by a five-bay gallery, in the middle of which is a going-through gate and on the east and west of which are the flights of four steps.

The three-bay back hall named Sansheng (Three Saints) Hall wears the flush gable roof of green glazed round tiles and yellow sheared edge. Various gods such as Taiyi, Tianqi (Emperor Dongyue) and Taiyin are enshrined. In front of the Sansheng Hall, a tablet named "Record of the Completion of the Hongci Palace" built in the 18th year under the reign of Emperor Wanli (1590). Behind the halls is a five-bay Rearmost Hall of two floors, wearing flush gable roof of green glazed round tiles and yellow sheared edge. On the first floor, four Heaven Generals are served as the guards and the Sanyuan Irritation Gods are enshrined; on the second floor, the Heavenly Gods whose statues are both delicate and simple are enshrined. On the left and right in front of the Rearmost Hall stand two shorea trees. According to *Notes of the Glass Street*— "the two shorea trees are too large for two persons to embrace. They provide a green environment and pleasant shade. They are also very fragrant. If one touches them, the fragrance will stay in one's hand for quite a long time."

Ever since Ming and Qing Dynasties, the Guangren Palace has opened from April 1st to 15th every year in lunar calendar. In Qing Dynasty, the government officials were sent to burn incense and worship. Until late Qing Dynasty, countless pilgrims still went to the temple to burn incense. The whole palace was grand. That the main hall and the Resting Hall are connected by galleries is the layout feature of the Song and Yuan Dynasties.

▲ The Rearmost Hall

# The Beiding Goddess Temple

As one of the Wuding Temples, the Beiding Goddess Temple (North Top Goddess Temple) is located in Deshengmenwai Tucheng, Chaoyang District. It is situated on the northern extended line of the south-north axis of Beijing and provides the important tangible historical materials for Beijing's development plan. It was designated by Beijing Municipal Government as a historical and cultural site to be protected in 2001.

The Beiding Goddess Temple was first built in the years under the reign of Emperor Xuande in Ming Dynasty. In Emperor Qianlong's reign in Qing Dynasty, it was rebuilt upon the imperial order. According to the historical data, there used to be a censer of the years under the reign of Emperor Wanli and a bell of the years under the reign of Emperor Xuande in the temple. Until the period of the Republic of China, there were over 40-bay halls, 120 statues and four stone tablets. In the following years, the temple suffered from several wars and became torn and tattered in 1950s. In 1976, the rear hall collapsed because of the earthquake and was later rebuilt into the Beiding

▲ The Main Gate

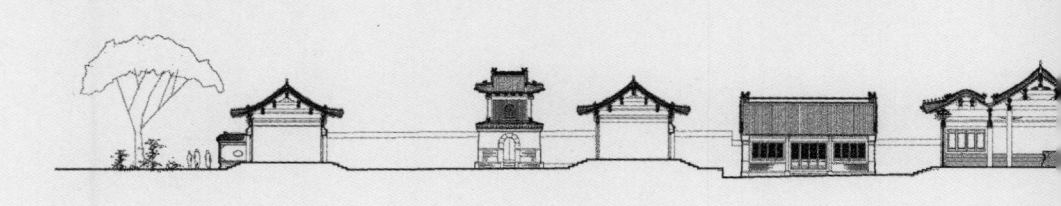

▲ Vertical Section of the Beiding Goddess Temple

Elementary School. In 2008, when Beijing held the Olympic Games, the National Aquatics Center "Water Cube" was constructed 100 meters north to its designed site so that the ancient buildings in the temple could be protected. The temple was also renovated at the same time so that the ancient buildings of Beiding Temple and the modern buildings of two new stadiums standing together to show Humanistic Olympic spirit.

The south-facing Beiding Goddess Temple originally has the following buildings from south to north: the main gate, the Hall of the Heavenly Kings, the Goddess Hall, the Dongyue Hall and the Yuhuang (Jade Emperor) Hall, four rows of courtyards in total. The main gate is of three bays, wearing hip and gable roof of gray round tiles with the tangent circle patterns under the roof. It also has a fire gate, a stone inscription board on which were written "Chi Jian Bei Ding Niang Niang Miao" (Beiding Goddess Temple Built upon the Imperial Order) imbedded in the arch gate and crossing lattice arch windows on the two wing frames. Besides, the stone in front of the arch gate is engraved with beautiful patterns. There is also a flight of three steps in the front. The main gate is connected with a splayed screen

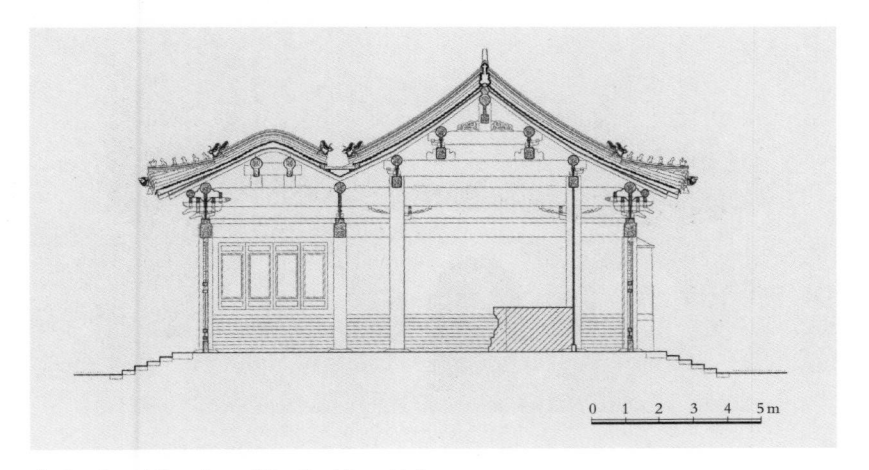

▲ Sectional Drawing of the Goddess Hall

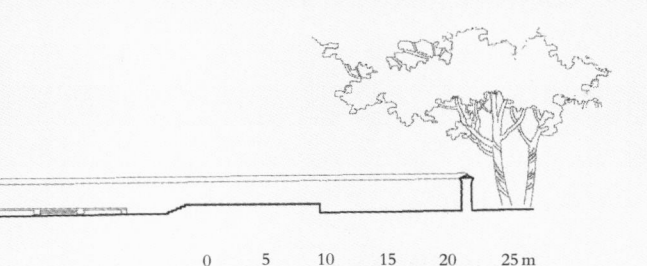

wall wearing hip and gable roof of gray round tiles. In the center of the screen wall are the patterns of flowers with four corners. Behind the main gate is the first row of courtyard wherein are the bell and drum towers wearing hip and gable roof of gray round tiles. On the first floor, there is an inward gate of fire with a flight of five steps in the front; on the second floor, arch windows are set in both the east and west sides. The Hall of the Heavenly Kings is of three bays, wearing hip and gable roof of gray round tiles. In the front wall, there are four partition gates; in the wing frames, there are four partition windows each with a flight of five steps; in the rear wall, there are four partition gates and the flanks of the hall are connected with partition walls to break up the inside from the outside. The Goddess Hall is five bays wide. In the front of the main hall is a three-bay extended veranda, wearing hip and gable rolling roof of green glazed round tiles and yellow sheared edge. The wing halls have flush gable roof of gray round tiles. Behind the Goddess Hall there are two more rows of courtyards with no buildings but the pedestal left. Besides, there are several stone tablets such as the table of "Record of Reconstruction of the Beiding Goddess Temple" written by Prince Yikuang in the 29th year under the reign of Emperor Guangxu(1903), three ancient Chinese junipers (with the oldest one of over 500 years) and five ancient Chinese scholar trees (around 300 years old).

▲ The Goddess Hall

# The Miaofeng Hill Goddess Temple and Lingguan Palace

The Bixia Yuanjun Temple on the Miaofeng Hill is located on the Golden Top, Jiangou Village, Mentougou District, also known as the Linggan Palace. Its main hall is enshrined five goddesses including Bixia Yuanjun Goddess, Yanguang (Eyesight) Goddess, Zisun (Offspring) Goddess, Banzhen (Macula) Goddess and Songzi (Children-Sending) Goddess, so it is commonly known as the Goddess Temple and was one of the most famous places where the temple fair is held. It was designated by Mentougou local government as a historical and cultural site to be protected in 1981.

There are four palaces and temples on the Miaofeng Hill: the Lingguan Palace on the hill passage, the Bixia Yuanjun Temple on the hilltop, the Huixiang Pavilion (also called the Tianqi Temple) on the northern hilltop and the Guandi Palace and Bodhisattva Palace in the hillside ten meters below the northeastern hilltop. Among them, the south-facing Bixia Yuanjun Temple is the most important one.

▲ Panoramic View of the Miaofeng Hill Bixia Yuanjun Temple

The Miaofeng Hill Bixia Yuanjun Temple was first built in the Ming Dynasty. At that time, the Taoist priest served as the abbot of the temple and the temple did not attract much pilgrims. In the 12th year under the reign of Emperor Kangxi (1673), it was named "North Top Tianxian Temple"; in the 28th year, Buddhism became more and more influential and the Buddhist became the abbot of the temple. Since then, it has attracted more and more pilgrims to come. In the 26th year under the reign of Emperor Qianlong (1716), it was reconstructed and renamed the Linggan Palace.

The main gate hall is of three bays, where is enshrined the Four Heavenly Gods. The main hall is of three bays with three inscriptional boards written by Empress Cixi, which mean "Light of Benevolence Shed on All Things", "The Merits and Virtues Equal to That of the Respected Woman", "The Auspicious Cloud Producing a Shade". On the niche sit the five female goddesses: Tianxian Bixia Goddess, Yanguang Mingmu Goddess (Great Eyesight Goddess), Banzhen Baohe Ciyou

▲ Tibetan-style Pagoda Covered with a Bowl-like Top

▲ The Main Gate Hall

▲ The Main Hall

Goddess (Protecting the Young Goddess), Zisun Guangsi Goddess (Many offsprings Goddess)and Songzi Xiqing Baochan Goddess(Guarding Child-Delivery Goddess).

In front of the temple is a six-meter high white marble pagoda in the shape of the bowl which was built in 1934. The Vajrapani, the sea and the cliff patterns engraved on the four sides of the square pedestal and the dancing dragon patterns engraved on the stone slab are very delicate and exquisite. There used to be east road and west road in the temple, where the Xishen Palace, the Fumo Palace and the Grandma Wangsan Palace, etc were situated.

On the Miaofeng Hill are enshrined gods of different religions, including Buddhist ones, Taoist ones and folk religious ones, which meet the various requirements of the pilgrims. Thus, it became such a religious holy land in the west suburbs of old Beijing that even Empress Dowager Cixi visited here to burn incense. At the time, there were four pilgrimage roads: the southern road Sanjiadian, the mid road Dajue Temple, the northern road Bei'an River and the old northern road Sha River. On the four roads, there are over a hundred of tea houses and temples as well as many service stations to provide assistance to the pilgrims for free. The pilgrims are not only from Beijing, but also from Tianjin, Fujian, Guangdong, Guangxi, Jilin, etc. The folklorists and sociologists from the early period of the Republic of China made much description and research on the Miaofeng Hill. Therefore, it is called "the cradle of Chinese folklore".

▲ Overlook of the Bixia Yuanjun Temple

# The Relics of the Yaji Hill Bixia Yuanjun Temple

The Yaji Hill is located to the west of Ru River, the north of Liu Jiadian County, Pinggu District with the altitude of 361 meters. It is not only a famous hill in Beijing but also a Taoist holy place in the Ming and Qing Dynasties. It was designated by Beijing Municipal Government as a historical and cultural site to be protected in 2001.

The Yaji Hill can be dated back the early Tang Dynasty when some Taoist priest wandered around here to settle and make pills of immortality and cultivate vital energy. In the 6th year of the Zhenguan's reign of the Tang Dynasty (632), a temple was built on the West Top, which was renamed the Bixia Yuanjun Temple in the Yuan Dynasty. During the years under the reign of Emperor Jiajing in the Ming Dynasty, a follower of Taoism commonly known as Grandma Wang'san in Xianghe County expressed her wish to rebuild the temple and collected iron tiles to build the Bixia Yuanjun Temple (known as Iron Tile Temple). In the 7th Tianqi year in the Ming Dynasty (1627), the province censor Ni Wenhuan asked Eunuch Wei Zhongxian to build a temple half way up the hill which was named

▲ Panoramic View of the Yaji Mountain

▲ The Bixia Yuanjun Temple

the Chonggong Temple (Worship to Merits Temple) by the emperor, yet it was not finished because Wei was convinced of guilt. During the years under the reign of Emperor Kangxi in the Qing Dynasty, the Yuhuang Pavilion was built on the Yaji Hill and the Huixiang Pavilion (Returning of Fragrance Pavilion) was built on the site where Wei built the temple previously. Afterwards, at the foot and on the half way of the hill, over 20 buildings such as the Xunshan Temple, the Sanguan Temple, the Bodhisattva Temple, the Dongyue Temple, the Lingguan Temple, the Guanyin Temple, the Chongwang Temple, the Zixiao Temple, the South Heavenly Gate and the Huye Temple, etc were built one after another and these are almost all the ancient buildings in Yaji Hill. The temples are built from the foot to the top of hill, arranged delicately. Since 1940s, most of the buildings in Yaji Hill have been demolished, only broken stone tablets and temple pedestals left. In 1980s, the Yaji Hill was developed as a tourist attraction. The local government raised funds to rebuild Xiding Bixia Yuanjun Temple, to renovate Zixiao Palace at the foot of the hill and to resume the folk temple fair. From 2003 to 2006, the temples in the Yaji Hill have been rebuilt basing on their original outlook.

The temples in the Yaji Hill can be divided into three groups:

On the top of the hill are two Yaji(bifurcating bob)-shaped rocks which are commonly known as East and West Top. They took shape by piled rocks around the hilltop with the height of over 30 meters. On the West Top, the main building is Bixia Yuanjun Temple where only the main three-bay hall was left, commonly known as the Goddess Temple. On the east of the temple stands a tablet on which were written "The Reconstruction of the Bixia Yuanjun Temple in the Yaji Hill"; on the north is the site of the Doumu Palace (Shengmu Pavilion). On the East Top is the Yuhuang Top whose size is half of that of the West Top. The Yuhuang Pavilion was originally built here, however, it was demolished later because part of the hill in the north collapsed due to the war. A cliff was therefore formed. Now only part of the pillar bases and cornerstones are left. In 2004, the collapsed part of the hill in the north was consolidated and repaired and rubble masonries were also added on the surface. The extant buildings in the Yuhuang Pavilion were restored in 2006. Going downstairs, visitors can see a newly-repaired Drum Tower. Between the two tops is the Sanhuang Palace with two royal tablets. On the south of the Sanhuang Palace is the main gate of the temple, outside which, there is the Wanshou Pavilion (Longevity Pavilion). Inside the pavilion is the tablet with the inscription written

▲ The Bell Tower

▲ The Yuhuang Pavilion

by Prince Cheng, He Shuo. The main gate of the temple is south-facing with Quemen (tall buildings on each side of the gate) on both the left and right sides. The main gate is connected with Quemen by the wallclamps. At present, only the platform over the ground, the stone template under the threshold, the intermediate pier and the drooping belt stone, etc are left. Based on the relics, it is judged that both the main gate and Quemen are of masonry structure. The main gate has hip and gable roof, while the Quemen has steamed bun gate. In 2006, according

▲ The Sanhuang Palace

Taoist Temples

233

to the relics and related records, buildings such as the Sanhuang Palace and main gate were resumed.

The main building in the middle of the hill is the Huixiang Pavilion, on the south of which are the Dongyue Temple and the Lingguan Palace; on the north of which are the Bodhisattva Temple, the Sanguan Temple and the Xunshan Temple, etc. On the site where the Huixiang Pavilion is, was formerly the Wei Zhongxian's Memorial Hall. Because of the splendid environment, the Taoist priest Li Juxiang built the Huixiang Pavilion here so that every April, pilgrims from Beijing and other areas around the country could pay a visit.

The extant building at the foot of the hill is the Zixiao Palace (also called the West Palace by local people) whose courtyards are the most complete among those temples that have suffered from the war. On the east of the Zixiao Palace is a three-bay main hall.

▲ The Huixiang Pavilion and the Dongyue Temple

Taoist Temples

▲ The Huixiang Pavilion

# The Fire God Taoist Temple

The Fire God Taoist Temple, also known as the Temple of the Perfect Sovereign of the Virtue of Fire, is located at No. 77, Di'anmenwai Street, Xicheng District of Beijing. Belonging to the Zheng Yi (Orthodox Unity) Taoist School, the large temple where the Fire God is enshrined was protected as a historical and cultural site of Beijing in 1984 by Beijing Municipal Government.

The Fire God is also called the Great Emperor of Zhenwu. Being one of the highest gods in Taoism, the Fire God is often mentioned in Taoist scriptures with several other names. Moreover, the god boasts some other nicknames among the ordinary people because of the popular and widespread folk belief in him. Meanwhile, the Fire God Taoist Temple, as an important ritual and ceremonial architecture, used to play an integral role in China's national sacrificial activities during Ming and Qing Dynasties. Since Ming Dynasty, whenever it came to the 22nd day of the June (lunar calendar), officials in charge of rituals would come to the temple and offer up sacrifices. The tradition did not cease until the period of the Republic of China.

The Fire God Taoist Temple, possibly built in the 6th year of the Zhenguan's reign during Tang Dynasty(632), was refaced in the 6th year of the Zhizheng's reign during Yuan Dynasty (1346). During its second renovation in the 33rd year of the Wanli's reign during Ming Dynasty(1605), the roof of the temple was retiled with green glazed tiles and two wings were added to the main temple. It underwent two more facelifts in Qing Dynasty, first in the reign of Emperor Shunzhi and later in the 24th year of the reign of Emperor Qianlong (1759). It was during these times that the roof was retiled with yellow glazed tiles. After the founding of People's Republic of China in 1949, the temple was once transformed into a hostel and residential compounds of the army. The vacation and restoration of the temple's historical and cultural relics began in 2001 with the following June witnessing a thorough and inclusive renovation of the temple.

Buildings of the south-facing temple are distributed along the north-south axis. In total, there are three courtyards. The east-facing main gate wears a hip roof with

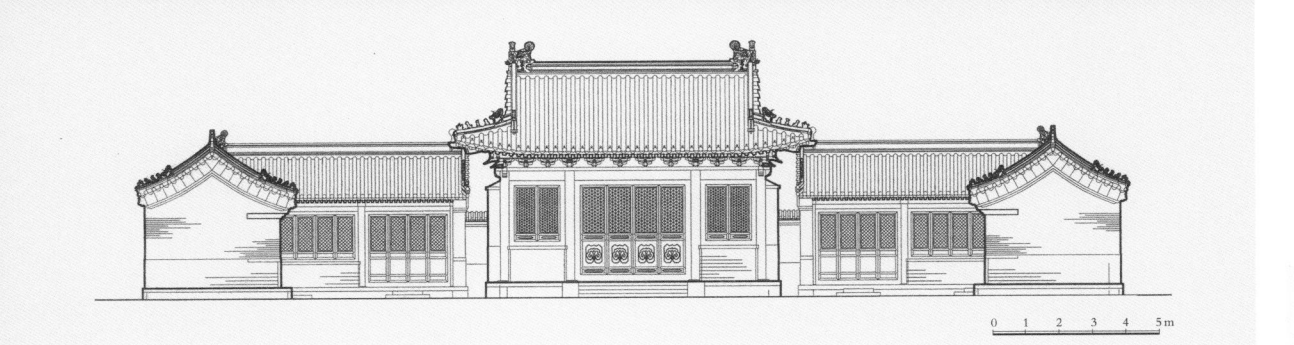

▲ North Elevation of the Long'en Hall and Corner Rooms

▲ The Fire God Taoist Temple

Taoist Temples

yellow glazed tiles and green cutting edges. Under the eaves is a bracket set of "one-arm-two-blocks" type. On each side of the main gate, there was a decorated gateway. The rebuilt gateway inside the main gate survives today, featuring four columns supporting its three-layered body. Other buildings inside the main gate are a bell tower and a drum tower. Three north-south courtyards are to the west of the main gate. Within the first courtyard, there are the South Hall, the Center Hall, and the West Wing Hall. The South Hall, named Long'en Hall (Hall of the Perfect Sovereign of Great Kindness), includes three rooms, wearing a gable and hip roof with black glazed tiles and green cutting edges. Under the eaves stands a bracket set with one lever and three horizontal bracket arms. The south wall of the hall is embedded with an inscription board reading "The Fire God Taoist Temple". The hall, which was used to enshrine Master Wang lingguan, the Heavenly General of the Jade Pivot Fire Office, is now transformed into a main gate. There are eight corner rooms next to the hall, four on each side, all of which are wearing flush gable roofs with green cutting edges. The Center Hall, known as the Fire Patriarch Hall , contains three bays and, in the front, a three-roomed

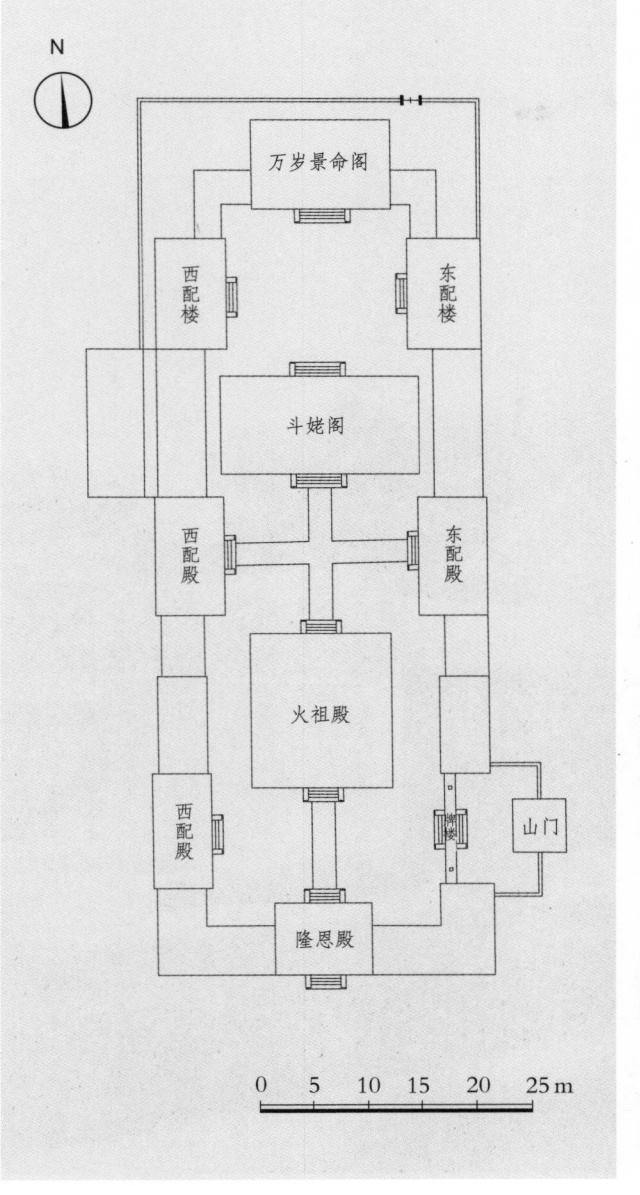

▲ General Layout of the Fire God Taoist Temple

▲ The Main Gate

annex with four purlins. The hall wears gable and hip roof with green cutting edges. The eight-sided wooden sunk panel in the hall, decorated with exquisite gilded coiled dragons, is one of the few well-preserved sunk panels extant in Beijing. The West Wing Hall also has three bays, but wears flush gable roof with black glazed tiles and green cutting edges. To the east and the west of the principal room, there are three small rooms on each side; all are wearing flush gable roof with black glazed tiles and green

Taoist Temples

▲ The Pailou

▲ The Fire Patriarch Hall

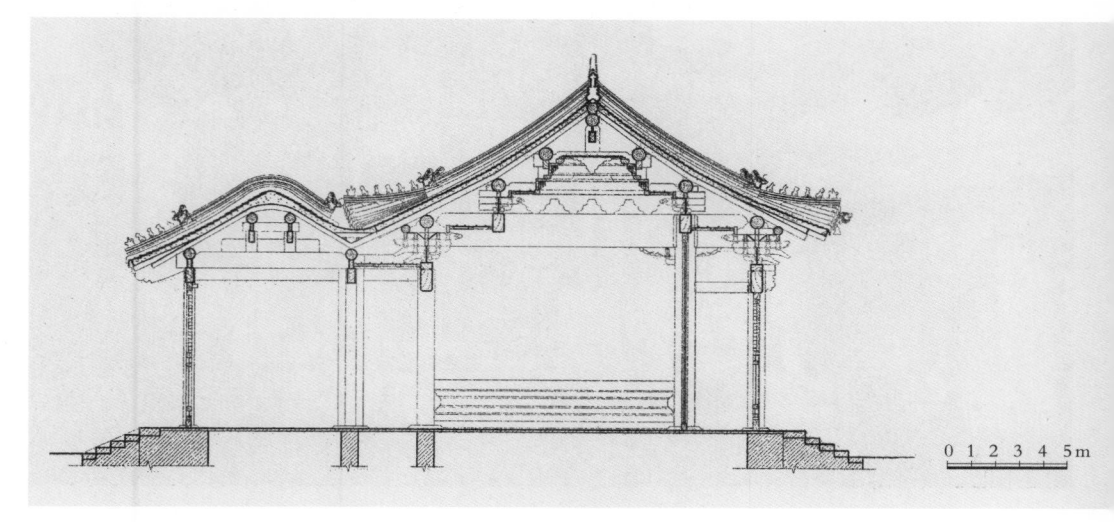

▲ Sectional Drawing of the Fire Patriarch Hall

cutting edges. Further north, there are two penthouses respectively on the east side and the west side, wearing flush gable roof and covered with round tiles. The Doumu Pavilion, the two-floored main hall of the second courtyard has five bays, wearing flush gable roof with black glazed tiles and green cutting edges. Under the eaves, there used to hang a board inscribed by Emperor Qianlong. The Wing Halls on the east and west each has three bays with colonnades in front, wearing flush gable roof with black glazed tiles and green cutting edges. In the third courtyard, there is again a two-floored building, commonly known as the Jade Emperor Hall. The hall is three bays wide wearing flush gable roof with yellow glazed tiles and green cutting edges. Under the eaves, there used to hang an inscription board by Emperor Qianlong. On the east and the west of the principal building are the East Wing Hall and the West Wing Hall, each with three bays, wearing flush gable roof with gray round tiles and green cutting edges. At the back of the temple, there used to be a water pavilion overlooking the beautiful sceneries of Shichahai Lake, yet it no longer exists.

▲ Sunk Panel of the Fire Patriarch Hall

▲ The Doumu Pavilion

▲ The Jade Emperor Hall

# The Xuanren Temple

The Xuanren (publicizing humanity) Temple is located at No.2 and No.4, Beichizi Street, Dongcheng District, Beijing. It is the only imperial temple in Beijing where the god of wind was enshrined in the Qing Dynasty. It is also called the Fengshen Temple (God of Wind Temple). Being one of the "eight temples outside the Forbidden City", it was designated by Beijing Municipal Government as a major historical and cultural site to be protected in 1984.

The Xuanren Temple was built on the imperial order in the 6th year under the reign of Emperor Yongzheng in the Qing Dynasty (1728) and was repaired in the 9th year under the reign of Emperor Jiaqing (1804). According to the *History of the Qing Dynasty*, the temple was constructed imitating the Shiying Palace (where the God of Dragon was enshrined) in Zhongnanhai. It was entitled by the emperor "Giving Protection in Time" and was named "Xuanren". After the temple was built, it was kept by the Taoist priests and financially supported by Imperial House. Every spring and autumn, the government sent officials to offer sacrifices. During the period of the Republic of China, religious activities of offering sacrifices to the god of wind were stopped and the temple became state-owned. The Taoist priests were all dismissed. Besides, organizations such as the Health Laboratory were established here. After the foundation of the People's Republic of China, the temple was used by the acupuncture clinic of Beijing Hospital of Traditional Chinese Medicine, the Retired Officials' Center of Beijing Municipal Bureau of Health successively. In 2003, the Xuanren Temple underwent large-scaled renovation.

The temple is south-facing with an east-facing street gate which was rebuilt after completion of the temple. The south-facing main gate is of three bays, wearing the single-eave hip and gable roof of yellow glazed tile and green sheared edges. Under the roof stands a bracket set with one lever and three horizontal bracket arms. Embedded in the stone arch door is an inscriptional board on which were written "Chi Jian Xuan Ren Miao" (Xuanren Temple Built upon the Imperial Order). The main gate is connected with a splayed screen wall. In front of the gate is a glazed brick screen wall wearing flush gable roof with a large ridge and of green glazed round tiles. It was built on a stone Sumeru base, embroidered with golden and exquisite flowers. Within the main gate are situated a bell tower and a drum tower. Having two stories, they wear hip and gable roofs of yellow glazed tiles and green sheared edges. On the first floor is the stone arch gate; on the second floor is the sun-block shutter. Under the eave is an arch with a protruding part in the shape of the head of the locust. The front hall is the Xian Hall (Obeisance Hall) where Uncle Wind is enshrined. It is of three bays wide, wearing single-eave hip and gable roof of yellow glazed tiles and green sheared edge and the timberframed color patterns. The main hall is the Xiang Hall (Sacrificial Hall) where Bafeng God is enshrined. It is of three bays, wearing a hip and gable roof with yellow glazed tiles and green sheared edges. Under the roof are five arches decorated with color patterns. The steps in front of the hall were made of white marble with dragons engraved, used only by royal families. The interior hall was decorated with exquisite flowers and flying dragons. The rear hall is the Resting Hall where the god of wind is enshrined. It is of five bays wide, wearing single-eave hip and gable roof of yellow glazed tiles and green sheared edge. The Resting Hall is flanked by two three-bay wing halls with flush and gable roofs of round tiles. In front of the hall is a front gallery. The west wing hall was renovated and the east wing hall is well preserved. In the temple there is another inscriptional board on which were written "Xie He Zhao Tai" (Harmony and Peace) by Emperor Yongzheng in the Qing Dynasty.

▲ Stone Board on the Main Gate

▲ The Resting Hall

▲ The Ceiling of the Resting Hall

# The Ninghe Temple

The Ninghe Temple(Cohesion and Harmony Temple)is located at No. 46, Beichizi Street, Dongcheng District. It was designated by Beijing Municipal Government as a historical and cultural site to be protected in 1984.

Built upon the imperial order in the 8th year under the reign of Emperor Yongzheng (1730), imitating the form of the Xuanren Temple, it is a temple where the god of cloud was enshrined. In the temple there is an inscriptional board on which were written "Xing Ze Zhao Cai" (Luster and Splendor) by Emperor Yongzheng. Since it was not far from the Forbidden City, many officials who came to Beijing to report their work or handle official business often lived here. During the period of the Republic of China, the temple served as a school and now it is the Beichizi Elementary School.

▲ Panoramic View of the Ninghe Temple

All the buildings inside the temple are south-facing, including the bell tower. the drum tower and four great halls. The temple gate is of three bays, 16.8 meters wide and 6.6 meters deep. It wears a single eave hip and gable roof of yellow glazed tiles and green sheared edge. Under the roof stands a bracket set of three arches decorated with color patterns. Both of the front and rear parts of the central room have arch gates, while the wing room has the stone carving arch windows. Inside the main gate, there is a bell tower and a drum tower. Both of them have two floors and take the shape of a square. They wear hip and gable roof of yellow glazed tile. The Xian Hall (Obeisance Hall) is of three bays wearing flush and gable roof of black glazed tiles and green sheared edge with a big ridge and the timberframed tangent circle patterns. The Xiang Hall (Sacrificial Hall) is of three bays wearing single-eave hip and gable roof of yellow glazed tiles and green sheared edge. The rear hall is of five bays with single eave hip and gable roof of yellow glazed tiles and green sheared edge. The rear hall is flanked by two wing halls of three bays wearing flush gable roof of round tiles with high ridge.

▲ The Imperial Stone Road in front of the Xiang Hall

▲ The Xian Hall

▲ The Xiang Hall

Taoist Temples

245

# The Sanjiadian Dragon King Temple

The Sanjiadian Dragon King Temple is located in the west of Sanjiadian Village, Mentougou District. It was built for the mother river of Beijing, Yongding River. With a relatively small size and exquisite architecture, it has been the best preserved building in Mentougou District. It was designated by Mentougou local government as a major historical and cultural site to be protected in 1981.

The south-facing Sanjiadian Dragon King Temple was first built in the Ming Dynasty. It took the form of the three-section courtyard. According to the inscriptions on the tablet in the temple, in the 14th year under the reign of Emperor Chongzhen in the Ming Dynasty (1641), Hou Yin came here and purchased the land of the neighborhood. Due to his diligent work in the field, it has changed from a wild field to a fertile land. Since the Yongding River was nearby, the land suffered from the river flood every year. Hou Yin thought the flood was the tricks played by the gods, so he raised donations and built the Longxing Convent. In the 51st year under the reign of Emperor Qianlong (1786), it was renovated and renamed "the Dragon King Temple". The temple has been renovated for several times afterwards.

The main gate of the Sanjiadian Dragon King Temple is of one bay. It wears the roof of round tiles and the high ridge. Under the roof is an inscriptional board on which are written "Ancient Dragon King Temple". In front of the gate are a pair of square gate piers and a pair of stone lions. The gate is connected with the walls on its flanks and a four-purlin cottage with round ridge roof of one bay. Under the roof are very delicate patterns and decorations. The main hall is of three bays with front gallery, wearing flush gable roof of round tiles. The rafter head under the roof is decorated with bronze plates of animal face patterns and Suzhou-style color paintings. While the pillars are decorated with sparrow braces which are different from those used in other temples. The sparrow brace in the other buildings are of carved flowers and grass patterns, while the one in this temple is flying dragon pattern (The golden dragons on the auspicious cloud are special and unique). Meanwhile, under the architrave are the spandrels in shape of downward-pointing triangles and hollow engraving auspicious clouds patterns. In the facade of the

▲ The Main Gate

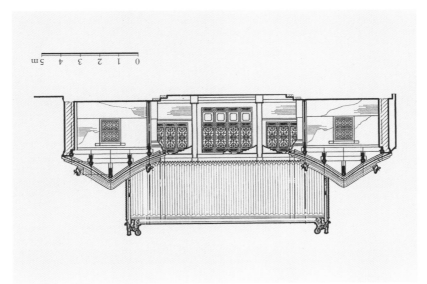

▲ Elevation of the Main Hall and Sectional Drawing of the Side Hall

▲ The Main Hall

temple, the central bay has the partition air door and the wing room has fan-shaped window. Both of them are exquisite. In the north of the temple are the wooden shrines with carved patterns. The middle three are one group and the one on each side is a group. The shrine contains five gods made in the years under the reign of Emperor Qianlong. The middle and west shrines contain the dragon kings from the four seas. In the east one is enshrined the God of Yongding River (the only extant statue of river god), which is rare among the Dragon King temples in Beijing. The statues are all magnificent with solemn looks. They are precious artistic works. On both sides of the middle shrine is a pair of couplet with black background and golden characters. The first line reads "The treasure god who safeguards the happiness of the people", and the second line reads "The fortune god who takes charge of food and clothes of the people". Behind the statues is a fresco of a great dragon flying in a cloud turning its head back, which is gorgeous. On both sides of the walls in the temple is the color painting "Dragon King Making a Trip" with a lot of characters such as thunder god, light goddess, rain god and wind god who are welcoming the Dragon King under a large flag. Among the auspicious clouds, there are dragon-driven carts. The painting is bright-colored and exquisite, which is the masterpiece of paintings in the folk temples. The east and west wing halls are of three bays, wearing

▲ Statue of the River God in the Main Hall

▲ Statues in the Main Hall

▲ Sparrow Brace of the Main Hall

flush gable roof of round tiles.

There are three stone tablets in the Sanjiadian Dragon King Temple, which are all under the roof of the front gallery in the main hall. In the second year under the reign of Emperor Shunzhi (1645), a tablet of the "Inscriptional Record of Reconstruction of the Longxing Convent" was set up. In the 51st year under the reign of Emperor Qianlong in the Qing Dynasty (1786), a tablet of the "Reconstruction of the Dragon King Temple" was set up. In the 17th year under the reign of Emperor Guangxu (1881), a tablet of the "Reconstruction of the Dragon King Temple" was set up.

Since the construction of the Sanjiadian Dragon King Temple, it has attracted a large number of pilgrims. In the Ming Dynasty, it was kept by the nuns. In the early Qing Dynasty, Buddhists took it over. On the 1st and 15th days of every month in lunar year, scriptures were chanted. In the period of the Republic of China, there were no Buddhists in the temple, but the Irrigation Association came here on the 1st, 15th day of every month, the Spring Festival and June 13th in lunar calendar (the birthday of the god of river) to pray and worship the Dragon King. This temple, therefore, has become the most popular one among the Dragon King temples in the west Beijing. For a long period of time, one of the major activities to worship the god of the river—celebrating its birthday is very characteristic and became a local custom. Every June 13th in lunar calendar is the birthday of the River God, villagers would offer sacrifices to the river gods and dragon kings in the Dragon King Temple. They lay Buddha amulets on the platform, make a feast, and eat longevity noodles to celebrate the birthday of River God, which is very special and unique.

Originated in the Holy Land Mecca, Islamism has been spread in Beijing for more than one thousand years. Consequently a number of Islamic mosques have been constructed in Beijing. Different from the Buddhist and Taoist temples built among mountains where few lived, Islamic mosques were all constructed in the communities where the Hui minority lived. Due to their long history and large scale, the Niujie Mosque, the Tongzhou Mosque, the Dongsi Mosque and the Huashi Mosque were renowned as the "Four Great Islamic Mosques". The four famous mosques are still serving as the venue for Muslim religious activities now.

Islamic Mosques

# The Niujie Mosque

Located at No. 88 Xuanwu District in Beijing, the Niujie Mosque is one of the largest mosques now extant. It was designated by the State Council as a major historical and cultural site to be protected in 1988.

The mosque was first built in the 14th year under the reign of Emperor Tonghe in Liao Dynasty (996), i.e. the 2nd year under the reign of Emperor Zhidao in Northern Song Dynasty. It was constructed by an Arabic scholar who served as an official in Liao Dynasty. It was renovated in the 7th year under the reign of Emperor Zhengtong in Ming Dynasty(1442) and was imperially named "Mosque" in the 10th year under the reign of Emperor Chenghua in Ming Dynasty(1474). In the 35th year under the reign of Emperor Kangxi in Qing Dynasty (1696), it was reconstructed as well as enlarged and the Emperor entitled it as "Mosque". During the Republic of China Period, the mosque was extended. In recent years, it has undergone major renovation and has become a place where Muslim people go to worship.

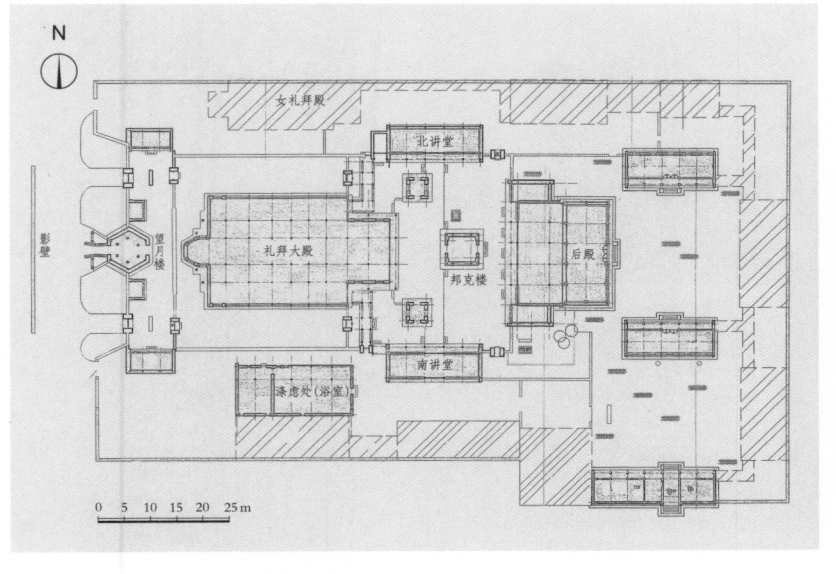

▲ General Layout of the Niujie Mosque

▲ The Screen Wall

Facing to the west Niujie Mosque is a typical architecture that combines Chinese traditional architectural style with Islamic-style decorations. The layout of the mosque is symmetrical and the main buildings are the Moon Watching Tower, the prayer hall, the minaret, the stele pavilion and the minbar.

Inside the entrance gate of the mosque is a large one-glyph-shaped screen wall of bricks and stones. Behind the wall stands a three-storied Pailou of four pillars, on which hangs an inscribed board saying "Broad Road to Heaven". Behind the Pailou is the Moon Watching Tower where Muslims stand to watch the moon. Being a hexagonal, two-storied structure, the Moon Watching Tower has a golden-glazed pyramid roof with green edge shear. Under the eaves are yellow tiles with green edges. A board saying "Niujie Mosque" hangs under the eaves.

Facing to the east, the prayer hall is the most important building in the mosque. Its front hall is three-bay wide. In the front aisle, pillars are used to support the eaves, under which there are bucket arches. Its rear hall is five-bay wide. The rear hall has smaller rooms aside. Together with the front hall, the total depth of the hall is 39 meters. It can house over a thousand prayers. The west-end of the hall is the kiln hall which signifies Mekka, the holy land of Islam. It is a hexagonal structure with a pyramid roof and bucket arches under the eave. In the hall the beam columns and ceilings are simply adorned with various paintings of ancient artifacts, flowers and Arabic scripts, which make the hall antique and elegant. In the north-western part of the hall is a wooden platform where sermons are given on Congregation Day

▲ Ridge Ornaments of the Screen Wall

▲ The Pailou and the Moon Watching Tower

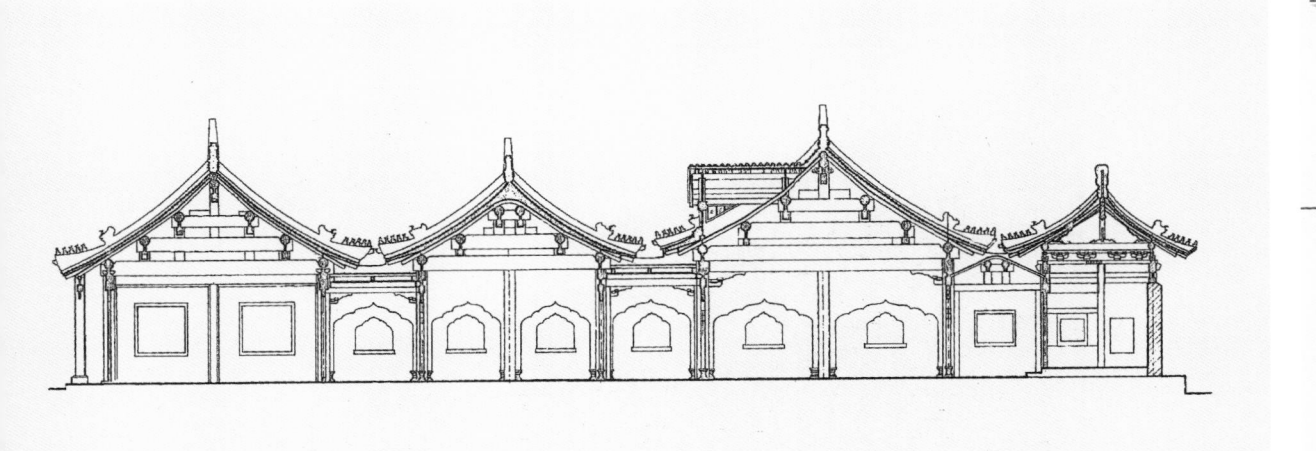

0 1 2 3 4 5m

▲ Sectional Drawing of the Prayer Hall

Islamic Mosques

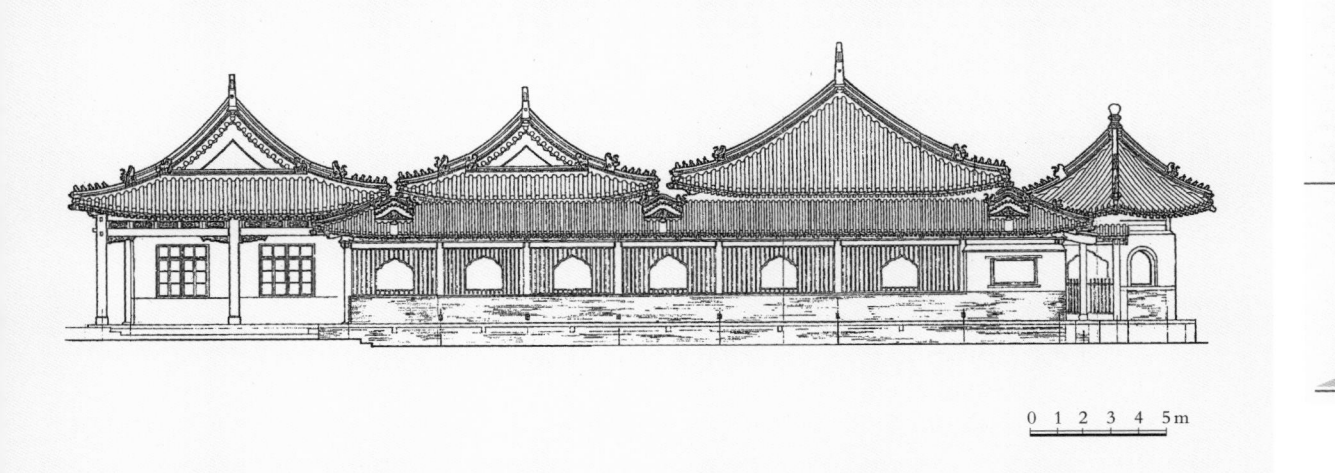

0 1 2 3 4 5m

▲ Side Elevation of the Prayer Hall

257

▲ The Prayer Hall

and other Islamic holidays.

The Islam ablution room to the south of the prayer hall was built during the period of the Republic of China. Being a western-styled structure it has an even-slope roof, brick-wall pillars, arched gates and windows. It was a place where Muslims washed their bodies clean.

The minaret is a square pavilion with a double-eaved gable and hip roof. It is a place for gathering the prayers and informing them the time to pray, so it is also named Xuanli (Adhan) Tower or Huanli (Adzan) Tower. The minaret was the former Scripture Storing Tower. In Kublai Khan years, two Arabic missioners had stored the Muslim scriptures there.

On both sides of the minaret, two square stele pavilions, each with a double-eaved gable and hip roof were built symmetrically. Two articles "On the Imperially Entitled Mosque" and "On the Reconstruction of the Imperially Entitled Mosque" were engraved both in Chinese and Arabic on the two steles in the 9th year of Ming

▲ Interior View of the Prayer Hall

Emperor Hongzhi's reign (1496) and 41st year of Ming Emperor Wanli's reign (1613).

On each side of the pavilion are five minbars connected to the prayer hall by verandas. East of the minaret are seven rear halls with hard top roof, front protruding veranda and pillars.

Besides, there are many other precious cultural relics which witnessed the spreading and developing of Islam in Beijing, such as the "White Plaque", gravestones in Arabic of Yuan Dynasty, a big copper pot, a copper censer, an iron censer, etc.

▲ Beam Frame of the Prayer Hall

▲ Inscribed Board of the Islam Ablution Room

▲ The Minaret

▲ The Stele Pavilion

▲ The Big Copper Pot

▲ The Rear Hall

Islamic Mosques

261

# The Tongzhou Mosque

The Tongzhou Mosque is located at No.1 Qingzhensi Street, Tongzhou District. It is one of the four great mosques in Beijing and was designated by Beijing Municipal Government as a major historical and cultural site to be protected in 1995.

The Tongzhou Mosque was first built in the years under the reign of Emperor Yanyou in Yuan Dynasty. It has a very long history, second to the Niujie Mosque. In the 11th year under the reign of Emperor Zhengde in the Ming Dynasty (1516), it was renovated and renamed "Chaozhen (Orienting Islam) Temple". During the years under the reign of Emperor Kangxi, Qianlong, Daoguang and Tongzhi in the Qing Dynasty, it underwent several expansions and renovations. In April, 1933, the Japanese troop bombarded Tongzhou County from the other side of the canal and destroyed some of its buildings which were repaired later. In 1945, a university funded by the mosque was established. In 1960s, several buildings such as the entrance, the screen wall, the Nanjing Pavilion and the Kiln Hall were all damaged. In

▲ Panoramic View of the Tongzhou Mosque

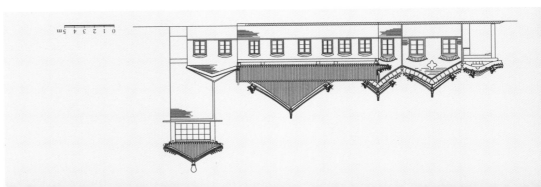

▲ Side Elevation of the Prayer Hall

1970s, it underwent a large scale of renovation and was reopened in 1996.

The east-facing mosque has the extant buildings of the prayer hall (north hall), the lecture hall, the hexagonal pyramidal roof pavilion, the minaret, etc. The east-facing prayer hall takes the form of connecting by four rows, 34 meters wide and 30 meters deep. The first and second rows are both three-bay wide with the former open hall and the latter gallery. The former wears the round turn-up roof and the latter the gable and hop roof. The third and fourth rows are of five-bay wide with the former wearing the roof of a big ridge and round tiles and the latter hip and gable roof of round tiles. Inside the hall, the foreshaft ceiling is decorated with peony patterns, while the tie beams all around are decorated with ancient utensils' patterns. Over ten hypostyle columns are all red-painted and surrounded by the splendid wires winding peony, which is quite unique. Behind the fourth row of

▲ The Prayer Hall

▲ Interior View of the Prayer Hall

the prayer hall is the Moon Watching Tower with a quadrangle pyramidal roof of glazed tiles. On both sides of the hills in the second row stand two Guomen Pavilions with hexagonal pyramidal roofs which are connected with the rear rooms in the third row. The minbar in the north of the main courtyard still a water room and five tablets mostly embedded in the wall. On the south and north of the mosque, there are tile-roofed houses connected with the mosque. The walls of the wing house are embedded with three tablets of delicate inscriptions.

▲ The Minaret

Islamic Mosques

265

# The Dongsi Mosque

The Dongsi Mosque is located at No. 13, Dongsi South Street, Dongcheng District. It is one of the four mosques in Beijing and was designated by Beijing Municipal Government as a historical and cultural site to be protected in 1984.

There are two opinions of the construction time of Dongsi Mosque. Some say that it was first built in the 6th year under the reign of Emperor Zhi Zheng in Yuan Dynasty (1346) by a sheikh's third son in East District of Beijing while some other people think it was first constructed in 12th year under the reign of Emperor Zhengtong in Ming Dynasty (1447) by a military officer named Chen You.

Facing east, the mosque covers an area of 4,000 square meters. It has three entrances, which were reconstructed into a hard tiled roof in 1920. Within the gates, there are several rooms. The gate combines the Chinese style with the western style. It is five-bay wide and has protruding verandas and brick front eaves. Originally a

▲ The Main Gate

▲ The Foyer

two-storied minaret with a squared quadrangular pyramid roof was built in the 22nd year of Ming Emperor Chenghua's reign (1486). But it was destroyed during Qing Emperor Guangxu's reign, the extant buildings are rebuilt in recent years.Behind the minaret, the major buildings inside are the prayer hall, the south and north minbars and the library. The prayer hall is on the west. It is the main building of the mosque, covering an area of 500 square meters. The front part of the hall is featured by the Chinese traditional pavilion wooden structure. There are 20 big colored pillars with the diameter of 48 centimeters, painted with big golden lotuses; On the three middle

▲ The Prayer Hall

beams are inscribed golden Koran. On each of the three arched gates, scriptures of Koran are engraved. On the southern end of the prayer hall, there is a stele which was set up in the 7th year under the reign of Emperor Wanli in Ming Dynasty (1579). With a name "Hundred-Word Praise of Muhammad", the stele is 91centimeters long and 67 centimeters wide and has a Sumeru base. The front of the stele is engraved the deeds of Muhammad who is the founder of the religion of Islam; the back of the stele is engraved "Infinity of the Rationale" in both Chinese and Arabic.

The mosque has always been one of the centers of Islamic cultural activities in Beijing. In 1926, an Islamic middle school was set up; in 1929, the mosque offered

▲ Interior View of the Prayer Hall

Islamic Mosques

269

▲ The Kiln Hall and its Dome

help to the Chengda Normal school which moved from Jinan, Shandong province to Peiping; in 1936, the mosque established the Fude Library; in 1947, it founded the Peiping Islamic School and published the Islamic academic and cultural journals such as *The Moon*, *The Muslims*, etc. Ever since the People's Republic of China was founded in 1949, the government has allocated funds to renovate the mosque for several times. Having Islamic Association of Beijing located inside, the mosque has been a gathering place for Muslims domestic and abroad to perform prayers, hold religious ceremonies and holiday celebrations.

▲ The Adornments in the Kiln Hall and its Dome

▲ The South Wing Hall

Islamic Mosques

271

# The Huashi Mosque

The Huashi Mosque is located at No. 30, Huashi Street, Chongwen District, Beijing. It is one of the four great mosques in Beijing and it was designated by Chongwen local government as a major historical and cultural site to be protected in 1984.

The mosque was first built in the 13th year under the reign of Emperor Yongle in Ming Dynasty (1415). It has a history of almost 600 years until now. Some say it was Changyuchun, one of the founding fathers of the Hui ethinic group in

▲ The Prayer Hall

▲ Inscribed Board inside the Prayer Hall

▲ Inscribed Board on the South Side of the Prayer Hall

Islamic Mosques

Ming Dynasty who transformed his mansion into the mosque, while some say it was Hu Dahai, the general of the Hui ethinic group who donated his own land for building the mosque. The mosque was rebuilt in the autumn of the 1st year under the reign of Emperor Chongzhen in Ming Dynasty (1628) and was expanded in the 41st year under the reign of Emperor Kangxi in the Qing Dynasty. In the 7th year under the reign of Emperor Yongzheng (1729), a royal pavilion was set up in which stands the imperial edict tablet issued by Emperor Yongzheng to protect the Hui ethnic group. In the 32nd year under the reign of Emperor Qianlong (1767), the neighboring fire brought disaster to the mosque which was rebuilt later. In the 25th year under the reign of Empeor Guangxu (1899), it underwent another renovation. After the founding of the People's Republic of China, the Democratic Management Committee was set up, in charge of the renovations of the mosque every year.

The mosque originally occupied the area east to Nanyangshikou, west to Yuchi Hutong, north to Xihuashi Street and south to Shoupa Hutong, which was larger than it is now. The extant buildings are all of Qing Dynasty style, including the pavilion, the Shigu (Learning from the Ancestors) Hall, the prayer hall and over ten subordinate houses in the north and south. The mosque gate faces the Huashi Street. The east-facing prayer hall is the main building. The front part of the hall is an open three-bay hall; the middle part is connected by three rows with flush gable roof. The hall has interior dome and on its roof are the structures of octagonal pavilions. The rear part of the hall is a kiln hall with a hexagonal pavilion, wearing square pyramidal roof. In front of the prayer hall is a royal pavilion with square double-eaved hip and gable roof. In the south, north and east of the courtyard are the minbar and other subordinate buildings. Besides, there are many tablets of renovation and recording from the Ming and Qing Dynasties and the period of the Republic of China in the mosque.

# POSTSCRIPT

*Beijing Ancient Architecture Series* is a large publishing project planned by the Beijing Ancient Architecture Research Institute and the Beijing Arts & Photography Publishing House attached to the Beijing Publishing Group. It was compiled by the Beijing Ancient Architecture Research Institute. This book is included as one of the volumes of this publishing project.

Hundreds of temples were preserved in Beijing. They have extremely rich connotations and important historic, artistic and scientific values. The task to show their values sufficiently increases the difficulty and pressure for the compiling work. With the full support of the leaders in the Beijing Municipal Bureau of Cultural Heritage, after the compiling staff underwent two years of arduous material collection and organization this volume was finally sent to the press.

The compiling work was greatly supported and assisted by the Culture and Heritage Commission of the related districts and counties, the Xiangshan Park, the Administration Office of the Yonghegong Lama Temple, the other administration organizations of the temples and the Beijing Arts and Photography Publishing House. With their enthusiastic help, we could smoothly complete all the drafts of this volume. My heartfelt gratitude goes to those who supported and assisted the compiling work of this volume.

In the process of compiling this book, Mr. Wang Shiren and Song Dachuan reviewed the content of this book closely and provided precious revision suggestions. Furthermore, in the process of writing this book, we also consulted Mr. Bao Shixuan and Mr. Wang Yi'e and both of them provided us with helpful instructions. My deep appreciation also goes to them.

At the meantime, the compiling of this book was supported by many photographers who offered a great number of high quality photos. A lot of photographers also actively participated in the pick-up shooting. We also own our thanks to them.

Besides, we cited and drew on many previous works that greatly help enrich the present book. Consequently, this book is the fruit of many colleagues who were engaged in researching and preserving the temples in Beijing.

Due to our own limitations, it is hard for us to avoid errors and omissions. We pledge the experts and readers to point out and correct our mistakes.

Editor